Economic and Fiscal Impacts
of Coal Development

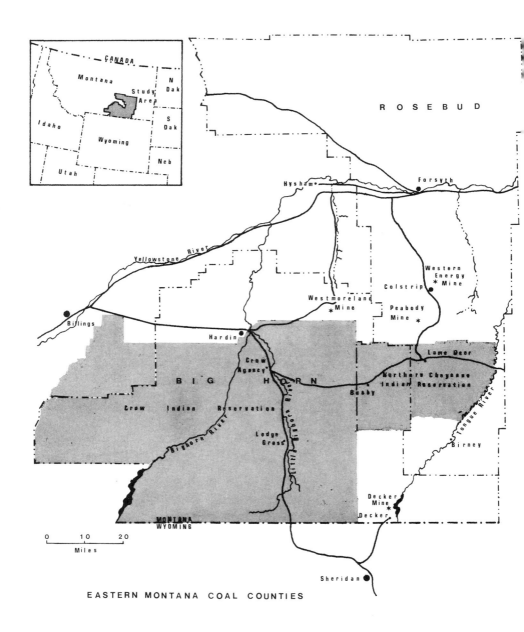

EASTERN MONTANA COAL COUNTIES

Economic and Fiscal Impacts of Coal Development:
Northern Great Plains

John V. Krutilla and Anthony C. Fisher
with Richard E. Rice

Published for Resources for the Future
by The Johns Hopkins University Press
Baltimore and London

Copyright © 1978 by The Johns Hopkins University Press
All rights reserved
Manufactured in the United States of America
Library of Congress Catalog Card Number 77-89300
ISBN 0-8018-2054-5
Library of Congress Cataloging in Publication Data will be found
on the last printed page of this book.

RESOURCES FOR THE FUTURE, INC.
1755 Massachusetts Avenue, N.W., Washington, D.C. 20036

2015456

Resources for the Future is a nonprofit organization for research and education in the development, conservation, and use of natural resources and the improvement of the quality of the environment. It was established in 1952 with the co-operation of the Ford Foundation. Grants for research are accepted from government and private sources only if they meet the conditions of a policy established by the Board of Directors of Resources for the Future. The policy states that RFF shall be solely responsible for the conduct of the research and free to make the research results available to the public. Part of the work of Resources for the Future is carried out by its resident staff; part is supported by grants to universities and other nonprofit organizations. Unless otherwise stated, interpretations and conclusions in RFF publications are those of the authors; the organization takes responsibility for the selection of significant subjects for study, the competence of the researchers, and their freedom of inquiry.

This book was begun in RFF's natural environments program and completed with a grant from RFF's quality of the environment division, directed by Walter O. Spofford, Jr. It was edited by Ruth B. Haas.

RFF editors: Joan R. Tron, Ruth B. Haas, Jo Hinkel, Sally A. Skillings

Contents

List of Figures

List of Tables

Preface

The oil embargo of the winter of 1973–74 brought with it the realization that the United States was vulnerable to decisions made by other countries regarding oil and gas supplies that could seriously compromise its ability to deal with foreign and domestic affairs. Among the strategies contemplated as a way to mitigate these circumstances was a greater reliance on domestic energy sources and technologies. Given the large reserves of coal in the United States, the substitution of coal for petroleum where possible (particularly for boiler fuel in steam electric generation) appeared to be a promising means of achieving a relative reduction in dependence on foreign supplies of oil and gas.

One hitherto largely neglected source of coal occurs in the Fort Union Formation in the Northern Great Plains. Indeed, the recoverable reserves in this formation are so large that they exceed the reserves of the northern and southern Appalachian coal fields combined. Moreover, they are relatively low in sulfur content, occur in extraordinarily thick seams near the surface, thus promising very low extraction costs, and are located in a region which, overall, has a somewhat higher annual precipitation, and more ample water supplies required for energy development than other more southerly western coal fields, as for example, the Four Corners area of the Southwest.

This study addresses the Montana portion of the Fort Union Formation—indeed, only the area represented by Big Horn and Rosebud counties, and certain peripheral areas of eastern Montana. Although the area might be considered small, it represents in our view one of critical importance. The relationship of eastern Montana coal fields to the potential electric utility demands of the northern tier of states penetrating as far east as Detroit and to the large interconnected power pool of the Pacific Northwest, with the latter's Northwest–Southwest intertie, suggests the pivotal location of these coal fields in an important set of regional markets. Under plausible conditions, this portion of eastern Montana could provide up to a quarter or more of the total increase in coal production called for over the next decade in the president's April 20, 1977 energy message.

The eastern Montana source of coal is doubly critical because local sentiment is not universally favorable toward large-scale coal development. Indeed, such development imposed upon a largely rural environment—a native culture found on the Indian reservations and the remaining stockman's subculture—is believed inevitably to bring in its wake profound changes in the region's ways of life. Moreover, the state of Montana, through its Utilities Siting Act, authority vested in its Natural Resources Board, and related environmental and health bodies, along with its ability to determine or challenge allocation of waters within its boundaries, is in a position to greatly influence the scale, character, and rapidity of energy development in this pivotal area.

For all of the above reasons, then, this area, which witnessed a confrontation between the "Feds" and the "locals" at the Little Big Horn almost exactly a century ago, is in another sense moving toward a confrontation over significant energy—and from the regional viewpoint, perhaps even more significant complementary water resources. These are issues which will affect the welfare of many Americans, not just Montanans.

This study focuses on this particular area and set of political jurisdictions because they will have a great deal of influence on the outcome of the conflicting claims and values. In any rapid, large-scale changes affecting the lives of people, there will be those whose welfare is improved and those whose welfare will suffer. There appear to be no completely value-free, objective standards by which to judge whether on balance the situation represents improvement or deterioration of the commonweal. We do not pretend to be able to render valid judgment regarding the merits of any outcome in such circumstances, but we are able to play out some of the quantitative implications of one or another policy option or energy development strategy.

By assessing a variety of policy-dependent economic, demographic, and fiscal outcomes quantitatively, we hope to provide information on at least some of the significant issues which have been raised in debate and thus reduce the degree of uncertainty concerning the implications of possible changes.

The proposal to employ a multiregional, multiindustry forecasting model developed by Professor Curtis C. Harris for investigating the economic and population impacts of different policy options originated with Anthony Fisher. Indeed, it was his task to review methods of simulating the implications of different coal development strategies in the Northern Great Plains, and evaluating their particular properties and characteristics. He is likewise principally responsible for chapter 2 of the manuscript. John Krutilla is principally responsible for chapter 1 and for chapters 3 through 9. Each, however, has also taken responsibility for reviewing and commenting on the draft materials that the other has prepared.

We have benefited from many sources in the preparation of this manuscript. Early in the formative stages of our research, we received a great deal of good counsel from John G. VanDerwalker, then Program Manager, Northern Great Plains Resources Program; from Don T. Nebeker, Associate Program Manager, Surface Environment and Mining Program, U.S. Forest Service; and Paul Meyers, Bureau of Land Management, Billings, Montana. Of unusually great service to us in both a facilitative and substantive role has been Theodore H. Clack, Staff Administrator, Montana Energy Advisory Council, who was ever responsive to our requests for information and to directing us to persons and sources who could be of greatest assistance to us. Indeed, without the active participation of John Clark, Administrator, Research Division, Department of Finance; Robert Stockton, Supervisor, State Aid Distribution, Department of Education; and Roger Tippy, Staff Attorney, Montana Legislative Council and their perceptive review and comments on early drafts of the study, our efforts would lack much of what merit they may ultimately possess. We have also benefited from assistance by John Fitzpatrick, Energy Planning Division, Department of Natural Resources; from Richard Dodge, Department of Community Affairs; Steve Turkiewicz, State Commission on Local Government; and from a review and comments on an earlier draft by Frank C. Montibeller, Jr., Administrative Officer, Montana Coal Board.

Numerous individuals at universities in the region have been of assistance to us. Professors Lloyd Bender and Anne Williams of Montana State University, Professor James A. Chalmers of Arizona State University and Mountain West Research, Inc., Professor Charles Howe of the University of Colorado, Professors I. James Pikl and William E. Morgan of the University of Wyoming, and Professors Maxine Johnson and Paul Polzin of the University of Montana have been the chief academic contributors to whatever success this study may enjoy.

We owe a special debt of gratitude to Sterling Brubaker of Resources for the Future, A. Myrick Freeman III of Bowdoin College and Resources for the Future, Thomas Muller of the Urban Institute, Kenneth Reifeld of the Bureau of Land Management, and Professor Thomas F. Stinson of the Economic Research Service and University of Minnesota for close readings and many especially valuable comments on an earlier draft of the study.

To Curtis C. Harris and Richard S. Davis, we are indebted for "fine tuning" the Harris model to the localized area under study and running the scenarios postulated to analyze the economic and demographic outcomes of various development options in Montana. Special mention needs to be made of the assistance Richard Rice and John Krutilla received from Connie Boris, Irving Hoch, and Kerry Smith, all of Resources for the Future, in attempting to deal with the issues addressed in chapter 8, and to Kerry Smith for assistance in formulating mathematically the tax, investment, and related computational models in the appendices to chapters 4, 7, and 8, which should facilitate use of the models by other investigators interested in different policy options.

Many persons have been from time to time most helpful over the two years this manuscript was intermittently in preparation, and their assistance may have been inadvertently overlooked in the acknowledgments. Joel Darmstadter, Jack Schanz, Toby Page, and Milton Russell of RFF narrowly escaped this fate. Doubtless there are others whose contributions have been unintentionally overlooked. To them we wish to express our appreciation, which is no less for their having been the victim of a lapse of memory.

Credit is due Rita Gromacki, Barbara Crawford, and Virginia Reid for their good-humored patience and perseverance in getting drafts of the manuscript into a form suitable for copy editing, and to Ruth Haas, whose talent for making manuscripts appear more handsome than they truly are and whose customary excellent editing goes a long way in making what we have to say available to whomever may be interested in it.

While Richard Rice appears on the title page, it is nonetheless appropriate to acknowledge that without his characteristic resourcefulness, diligence, and reliability, the study literally would not have been completed in any timely fashion.

J.V.K.
A.C.F.

January 1978

Economic and Fiscal Impacts
of Coal Development

1 The Energy Situation in Review

INTRODUCTION

The 1973–74 winter oil embargo may well have been a propitious and timely event. A trend to greater reliance on foreign sources had resulted in imports accounting for 35 percent of U.S. domestic petroleum consumption, and Middle East sources alone (including Iran) for 12 percent.[1] Should this trend have continued for another decade, it is likely that the United States would by then have been importing nearly 60 percent of its petroleum supplies, with the bulk of this coming from the Middle East. The producing countries would, given this dependence, be in a position to seriously disrupt the U.S. economy by sudden and violent changes in oil prices, or even a cutoff of supplies. The fact that a group of producers could organize a partial embargo in 1973 established the credibility of that threat at a time when its persuasive demonstration did not yet spell disaster for the U.S. economy.

For a nation that had been both the largest producer of petroleum and itself self-sufficient until the late 1940s, what factors accounted for the change in circumstances by the winter of 1973–74? One of the reasons domestic demand for petroleum exceeds domestic production relates to the reduction of sulfur emissions required by the Clean Air Act

[1] See Joel Darmstadter and Hans Landsberg, "The Economic Background of the Oil Crisis," *Daedalus* vol. 104, no. 4 (Fall 1975) pp. 15–37.

1

of 1967 and subsequent amendments. Conversion to low-sulfur oil and even cleaner natural gas was one means by which to comply with emission standards, although it was not the only one. However, it significantly increased the demand for petroleum. For example, in 1970 at a time when the use of petroleum by electric utilities was around 900,000 barrels a day, projected petroleum use for this purpose suggested a three-fold increase by 1985, in spite of the fact that nuclear power was projected to be a major component of new electric utility generating capacity.[2]

At the same time that the substitution of oil for coal by electric utilities in the United States was leading to an increase in the demand for petroleum, the yield of oil from domestic fields that could be supplied at historical prices of around $4 per barrel began declining. Accordingly, with increasing demands and declining yields from domestic oil fields, the resort to low cost (about $2 per barrel) foreign sources of supply appeared economically compelling.[3] Both the embargo imposed during the winter of 1973–74 and the subsequent quintupling of price by the Organization of Petroleum Exporting Countries (OPEC), however, have prompted reconsideration of the matter from strategic and economic perspectives. In short, a program designed to secure a substantial measure of independence from insecure foreign sources continues to be actively considered. The following sections present some of the options, their prospects and limitations, in an effort to evaluate the role that coal, and especially coal from the Northern Great Plains, might play in alleviating dependence on foreign petroleum sources.

We begin by evaluating the promises and limitations of energy conservation. Looking then at supply alternatives, we assess the role of nuclear fission, unconventional sources of energy, including solar and geothermal, and finally domestic fossil fuel sources. We do not address the problem of the long-term consistency between the stock of energy materials and consumption, that is, the need for a technology that will ultimately convert inexhaustible sources of energy, as exemplified by fusion or direct solar conversion. Nor do we consider the problem posed by energy materials that have cumulative environmental hazards and thus

[2] National Petroleum Council, *U.S. Energy Outlook: Nuclear Energy Availability.*

Of course the electric utility industry acounts for only a relatively small portion of total U.S. petroleum consumption, currently running at about 16 million barrels a day. But we focus on this use because there are good substitutes for oil in electricity generation, and not (yet, at least) in other uses such as transportation and the production of petrochemicals.

[3] The transition to imported oil was in fact slowed by the quota system that was put in place during the 1950s and disintegrated only slowly.

have ultimate limits to their conversion. Our concern in this study is to address the short-run problem—the prospects for adjustments in sources of energy supply over the next decade or so. At the same time, we shall be mindful of the environmental consequences associated with any short-run option adopted, especially where these are cumulative or difficult to reverse.

ENERGY CONSERVATION AND RESIDUAL DEMAND

In the United States, energy commodities and services historically have been priced substantially below their full social costs.[4] A national belief that intensive use of energy was a *sine qua non* of economic development led to preferential tax treatment for energy and mineral commodities and subsidy of energy commodities and related services, such as transmission, by public agencies. But perhaps the most important explanation for the fact that energy commodities have been priced below their social costs has been the failure, until recently, to recognize environmental damages as part of these costs. These damages appear variously as health costs to the general public when either industrial or automotive emissions are not regulated adequately; as reduction in the future productivity of the land when, for example, acid mine wastes enter streams, or sediments are deposited from eroding spoils banks associated with mining activities; or as landscapes disfigured by extraction activities. Such adverse third-party effects historically have not entered into the market costs at which energy commodities and services have been exchanged.

Because conventional market prices have not included environmental costs, users have historically consumed energy for some purposes in which the value of the energy was less than its social cost. Economic theory suggests that to achieve social efficiency in energy use, consumption patterns ought to be altered to bring value in use at the margin in line with social costs. However, in the evolving patterns of use based on underpriced energy commodities and services, complementary productive factors and consumption goods were linked to the price of energy. Where this has involved capital expenditures for plant, equipment, or durable consumer goods, production modes and consumer behavior patterns have become established for the lifetime of the equipment. That is, given the longevity of capital plant, equipment, and consumer durable goods which have reflected efficient combinations with energy only because of its historical

---

[4] This is true even if we ignore the so-called "user cost," or shadow price that attaches to the consumption of a depletable resource like petroleum.

underpricing, these capital investments now represent sunk costs that can be recovered only through continued use of such facilities. Only in rare cases would it pay, from either a private or social point of view, to abandon this sunk capital in favor of more energy-efficient production facilities or consumer durables. Note that even if energy had been "correctly" priced from the social point of view, users would still be locked into the existing plant and equipment. An increase in energy prices or even the discovery of a new, cheaper method of energy conversion would not ordinarily indicate the premature abandonment of the existing stock of capital.

This durability of sunk capital represents a problem in the short run when one looks to energy conservation as a means of substantially reducing the demand for energy commodities. Makeshift measures that encourage marginal behavioral changes, such as a lowered freeway speed limit that permits the operation of energy-inefficient vehicles at a more energy-efficient speed, promise some effective conservation in the short run. But the replacement of energy-inefficient equipment with a more efficient stock of vehicles or transport modes is a matter of grave difficulty in the short run because of the transition time involved. When we consider plant and equipment of greater durability, such as buildings, we appreciate that perhaps a quarter to a half century would be required to make the adjustment in our total stock of capital.

While the long-run promise of greater efficiency in the utilization of energy is extremely important, in the shorter run, say the next ten to fifteen years, the relative reduction in demand that can be achieved without serious economic disruption is somewhat more limited. Indeed, in the Ford Energy Policy Project Report, the time path of energy consumption for a "zero energy growth" scenario does not depart significantly from an alternative "technical fix" scenario until after fifteen or twenty years.[5] The latter is also an energy conservation scenario with a relative reduction in the annual growth rate from 4½ to 2 percent per year. And, realistically speaking, it is not even very likely that the reduction in growth to only 2 percent per year will be attainable within the next ten years unless energy conservation becomes a high priority national objective with more effective congressional and executive branch action than has yet become evident.

Given the 1973 energy consumption of 75 quadrillion British thermal units (Quads), growth at the recently experienced (1960–70) rate of 4½

[5] *A Time to Choose: America's Energy Future.*

percent per year would add another 52 Quads by 1985. Unless economic stagnation of long duration with much idled industrial capacity and unemployment occurs, an increment in energy utilization of about 30 Quads by 1985 appears to be about as much as one can realistically expect. Even assuming the most effective energy conservation program, namely, one that limits growth to about 2 percent per year, an additional 20 Quads would still be demanded annually by 1985. This is equivalent to roughly an additional 10 million barrels a day of petroleum—an amount slightly greater than our current domestic production. Alternatively, it would represent approximately 800 million tons per year of additional coal, or about one and one third times our current rate of production. In short, even taking the most optimistic view regarding the vigor and seriousness with which a national energy conservation policy is pursued, some expansion of energy supplies appears to be required to support a fully employed domestic economy over the next decade, at least.

A POTENTIAL NONFOSSIL FUEL OPTION: NUCLEAR FISSION

If oil is to be reserved for the transportation sector and petrochemicals, and we ignore coal for the moment, is there a nonfossil fuel option that would allow the displacement of oil (and gas) currently being burned as fuel in the production of electricity? This is technically feasible with nuclear fission reactors. Indeed, if we are to accept the Energy Policy Project's most conservative postulated rate of growth in energy consumption, namely the 2 percent per year, the projected rate of growth of new nuclear capacity, until recently, seemed sufficient both to displace oil requirements for electricity generation and to cover the projected nuclear share of generating capacity. It has become apparent, however, that the schedule for deploying nuclear power technology has lagged badly. Technical problems, including those related to reactor safety, have intervened to the extent that 1985 nuclear power capacity is expected to fall short by some 100,000–250,000 megawatts (MW) of the capacity projected only a few years ago.[6] Given the required lead time for a nuclear power reactor, there is now no way that we can realistically anticipate more than about 180,000 MW of nuclear capacity by 1985, and that will not be sufficient to cover its projected share of future generating capacity, even under the slimmest growth rate postulated, and still displace fossil fuels.

---

[6] Compare National Petroleum Council, *U.S. Energy Outlook: Nuclear Energy Availability* and *Nuclear News,* February 1974.

But aside from their inability to displace fossil fuels in thermal electric stations in time to deal with the short-term energy crisis, there is some question of the ultimate desirability of nuclear *fission* reactors in any event. While the at-site environmental effects of a properly functioning nuclear reactor may be less damaging than those resulting from the combustion of fossil fuels, particularly coal, nuclear fission does produce highly toxic radioactive wastes. Since these wastes, particularly the actinides associated with the liquid-metal fast breeder reactor, are very long lived (the half-life of plutonium, for example, is 24,000 years) there is need for failsafe storage of such wastes for very long periods. This poses at least two problems, one technical and one social.

The technical problem is to devise the storage facility. It now appears that commercial nuclear wastes will ultimately be stored as ceramic materials in deep, geologically stable salt formations in the Southwest. Although not everyone is persuaded that this will prove both feasible and acceptably safe, it seems to us less troubling than the social issue, which is how to assure a world order of the requisite discipline and stability to safeguard these materials. A closely related issue is how to safeguard materials, notably plutonium, that would be increasingly in use, transit, and storage as a result of a commitment to fuel reprocessing or the breeder reactor. In other words, the problems posed by storage of long-lived toxic by-products of fission and the handling of plutonium in reprocessing facilities are not just those of malfunctioning technical systems. They relate also to malfunctioning members of society bent on theft, sabotage, and related terrorist activities.[7] This in turn has serious implications for the kind of society that would emerge in response to the need to deal with them. Partly in response to these implications, interest has turned to a number of hopefully more benign, if still exotic, sources of energy.

NONFOSSIL FUEL, NONNUCLEAR OPTIONS

Solar

Most of the serious attention directed to displacement of petroleum in nontransportation energy uses has concerned coal and nuclear materials as substitutes. Yet, solar and geothermal sources of energy may be possibilities in heating residences and in generating electricity, even perhaps to some extent in the short run.

[7] See Mason Willrich and Theodore Taylor, *Nuclear Theft: Risks and Safeguards.*

A distinction should be made between lower and higher temperature uses of solar energy. Solar is basically a diffuse source of energy. The technology for its application in low-temperature uses such as space heating is simple and proven. In much of the country it has been potentially competitive, even at the former lower level of fuel prices, with other sources of energy for supplementary residential heating for some time, assuming new construction is designed to accommodate rooftop solar heating units. With the rising cost of fossil fuel, substantial displacement of such fuels in residential heating and possibly cooling could be accomplished with known technology. A problem, of course, involves the need for redesign or replacement of existing residential structures built without consideration of features essential to the efficient interception and utilization of the sun's rays. Replacement of the present housing stock would be a long process, and it is not clear how much of it could be retrofitted to use solar energy.

Because of the high capital cost in setting up a supplementary solar collector, it would take a decade or two to realize savings on the total cost of production. The reluctance of builders not only to pioneer a largely untried technology but in addition to substitute higher capital costs (more visible to buyers) for lower operating costs appears to be one of the main obstacles to installing solar space heating and cooling equipment.

The use of solar energy directly in high-temperature applications, on the other hand, does not appear immediately promising. Because solar energy is so diffuse a heat source, the investment in collectors that would concentrate sufficient heat to produce the temperatures required in conventional steam cycle electrical generating plants will be several times higher than the investment for fossil fuel and nuclear stations. Accordingly, the extensive use of capital as well as land for the higher temperature applications using direct insolation as an energy source make it appear a more dubious proposition.

More sophisticated technologies, such as the photovoltaic cell, which are capable of directly converting solar to electrical energy, are still further from commercial feasibility. Here, while the technology of the photovoltaic cell is known, present efficiencies in the economic sense are so low that it affords an energy source only in space exploration and similar highly specialized purposes where the cost per unit of electricity is an incidental consideration. The prospects for achieving the advances that would permit deploying a technology for producing moderately economical electricity from solar radiation do not appear promising much before the end of the century.

Geothermal

Geothermal energy is another potential substitute for nuclear or fossil fuels in the production of electricity. Heat sources in the earth's crust exist in the form of dry steam, hot water, and hot, dry rock. The Geysers field in Northern California is one of the large dry steam fields now producing electricity in the United States and is undergoing additional development. It produces steam at pressures and temperatures that are low compared with modern steam electric plants, but it has been suggested that with appropriate plant design, economical energy can be generated, depending on various factors, the life of the field being one. One estimate of the potential of dry steam sources in the United States places the amount at 100,000 MW over a twenty-year production cycle.[8]

Dry steam wells produce steam unaccompanied by liquid water, which makes power generation both more simple and economical. Geothermal wells more commonly produce a mixture of steam and water and the water must be separated, in the conventional process, from the steam before the latter can be used in power production. Where the water has a high salinity content, the disposal can create difficulties if it cannot be reinjected into the system after the steam has been extracted.[9]

Up to the present, geothermal technology has concentrated on dry steam and hot water systems. But the preponderant source of geothermal energy in the earth's interior is contained in dry rock. While the thermal content of dry rock is much greater, it is more difficult to exploit for the production of electricity. Plans to tap this source of energy involve the creation of artificial fractures in the rock at a substantial depth below the surface, and circulation of water through the area to extract the heat from the rock. Research is being conducted in this area, but there is as yet little empirical, contrasted with theoretical, information on which to base judgments regarding the comparative economics of this source. Were it to become cost-competitive, the dry rock source would represent a reservoir of heat substantially greater than the dry steam and hot water sources combined.

Although geothermal energy has come to represent a significant potential source of energy for the future, it is not a short-run solution to the energy problem. It is clear that much research and development is required if it is to be efficiently exploited. Moreover, it is recognized that

---

[8] U.S. Senate, Committee on Interior and Insular Affairs, *Geothermal Energy in the U.S.*, in Committee Print 92-31, 92 Cong. 2 sess., May 1972.
[9] Allen L. Hammond, William D. Metz, and Thomas H. Maugh II, *Energy and the Future.*

there are potential environmental problems. The mineralized nature of the water, and even residues in the dry steam, represent by-products of the process that require disposal. Where they may be reinjected into the reservoirs conveniently, they may result in no environmental problem and serve, as well, to prevent land subsidence that can pose a problem as steam and water are withdrawn. In other cases, however, there will be environmental costs that cannot be ignored.

Other Sources

Other potential energy sources include winds and tides, solid waste, and nuclear fusion. We are not aware of any estimate that indicates that any of these can account for a nonnegligible fraction of U.S. or world energy consumption in the short to medium term—say to 1985. Fusion does hold the promise of being the ultimate—inexhaustible, low polluting—source, but the consensus is that if it turns out to be feasible at all, it surely cannot even begin to make a contribution, outside the laboratory, until early in the next century. The implication of these remarks is not that nonfossil, nonfission sources should be ignored. On the contrary, over the very long run they clearly offer an alternative to dependence on limited stocks of fossil fuels, and fissionable uranium. When one considers that they also tend to do less damage to the environment, the case for accelerated research and development becomes overwhelming. But none of this changes the fact that, at least to the end of the century, fossil fuels are likely to be dominant. It is to prospects for domestic development of these resources that we now turn.

FOSSIL FUEL ENERGY SUPPLY OPTIONS

Just as the rise in the domestic prices of energy commodities is expected to have an impact on the quantity of fuels demanded, so is it likely to stimulate some response in supplies of fossil fuels. A near trebling of the domestic price of crude oil, which has taken place for new sources, will make some deposits that were marginal to uneconomic at the earlier price, now economically recoverable. How much of this will be available from onshore sources, including tertiary recovery, is not presently known. Since most of the new production has been offshore in recent times, it is likely that the offshore sources will remain the most attractive from the standpoint of private resource costs. As the environmental impacts of these offshore operations are currently under investigation, we do not yet have a means of determining whether the total, environmental as well as financial, costs affect the relative attractiveness of the offshore compared with onshore sources[10]—nor the supplies that will be forthcoming.[11] These are matters that require serious attention before we can schedule expanded production with the assurance that we are minimizing total social costs of domestic energy supplies.

Rising prices of petroleum products have implications also for the commercial feasibility of extracting oil from shale. A rough estimate of shale oil resources has been put at 168 trillion barrels,[12] with the richest zones of the Green River Formation of Colorado, Utah, and Wyoming yielding upwards of 25 gallons per ton, set at nearly 600 billion barrels.[13] If only private resource costs of extraction and transportation are considered, it is possible that these higher grade oil shale occurrences might be recoverable at 1975 prices, though even this is not certain. Rising costs of commercial inputs, particularly in construction, have kept estimates of oil shale costs moving upward. As a result, oil shale's relative economic position may well have remained unchanged.

The environmental costs of these deposits, however, are formidable. The Piceance Creek Basin in Colorado has an unusually important deer herd and related wildlife values. But destruction of these values probably represents the least significant of the environmental impacts of extractive operations for the regional environment. The mind-boggling volume of spoils, the attendant risk of disposal pile failure, the environmental implications of dewatering the mines, and related water quality problems, raise serious questions regarding the real costs associated with these sources.[14] Conceivably, processes may be developed that would eliminate

---

[10] A beginning in this direction has been made by the Council on Environmental Quality. See *OCS Oil and Gas—An Environmental Assessment* (Washington, April 1974).

[11] It is significant that the Project Independence Report suggests that at even an $11 price per barrel permitting secondary and tertiary recovery, production from already discovered oil reserves would scarcely remain equal to 1974 production. The increment in output over 1974 of an estimated 50 percent would come about equally from Alaskan North Slope and both Atlantic and Pacific Outer Continental Shelf sources.

[12] Donald Duncan and Vernon Swanson, "Organic-Rich Shale of the United States and World Land Areas," U.S. Geological Survey Circular 523.

[13] U.S. Department of Interior, *Propects for Oil Shale Development—Colorado, Utah, Wyoming.* By comparison, the consumption of oil in the United States in 1974 was approximately 6 billion barrels. Proved reserves, as of 1968, stood at no more than 40 billion barrels (see Joel Darmstadter, Perry Teitelbaum, and Jaroslav G. Polach, *Energy in the World Economy: A Statistical Review of Trends in Output, Trade, and Consumption Since 1925*).

[14] At 25 gallons to the ton, about 10 billion tons of spoil materials would be generated annually to produce oil at the rate of the 1974 U.S. consumption. U.S. Department of the Interior, *Final Environmental Statement for the Prototype Oil Shale Leasing Program, vol. 1, Regional Impacts of Oil Shale.*

a large share of the expected environmental cost.[15] Work is currently in progress that may eventually permit a careful evaluation of the social cost of environmental degradation inherent in the extraction of oil from shale. At the moment, however, partly because of the expected environmental component of cost, shale oil does not appear to be a strong candidate for the more economic among the energy supply options.

An alternative fossil fuel option would be to return to coal, which the United States has in relative abundance, to generate electricity coupled with effective emission control. In 1973, approximately 1.5 million barrels of petroleum were used daily by the electric utility industry. By 1985 the amount projected in 1973 by the National Petroleum Council runs, as noted earlier, to 2.7 million barrels a day. The 1973 annual consumption of petroleum used in electric generating plants was roughly equivalent to 135 million tons of coal, and this volume would almost double to 240 million tons by 1985. And if coal were to substitute for natural gas as well, between 350 million and 400 million tons of additional coal—or about two thirds of current production—would be required (Project Independence Report).

Initially, coal from low sulfur deposits could be substituted for low sulfur petroleum. Given time, and a vigorous research and development program to develop effective control technology for sulfur and particulate emissions, more extensive sources of coal could be drawn upon to provide the needed fuel. But the environmental impact of coal will be felt at the extraction stage as well as during conversion. Deep-mined coal has traditionally been attended by extremely high occupational hazards. The mining of subsurface deposits of coal is perhaps the most hazardous occupation in America, quite apart from harm to miners' health resulting from long exposure to airborne particulates. Advances in underground mining technology and their more widespread deployment, of course, could reduce the number of miners exposed for any given volume of coal mined.

Surface-mined coal is not occupationally so hazardous, but it has long-run landscape and related aesthetic costs associated with it. Where seams are thin and in areas of steep slopes, contour mining is particularly damaging to the landscape. In certain regions of the West, the Northern Great Plains, for example, where coal seams are exceptionally thick, and

---

[15] See, for example, Ben Weichman, "Energy and Environmental Impact from the Development of Oil Shale and Associated Minerals" (paper presented at the 65th Annual Meeting of the American Institute of Chemical Engineers, November 1972, revised 1974), for an interesting possibility in this connection.

the amount of surface area disturbed per unit of coal extracted is small (11 acres per million tons of coal, assuming an average 50-foot seam), the ratio of surface disturbance to amount of coal taken is much less. In parts of this region where annual precipitation is sufficient, it appears feasible to restore the strip-mined areas.[16]

It is felt by some that the surface mining of western coal, if carefully supervised and restricted to areas that can be restored, affords one of the least environmentally disruptive of the extensive sources of new energy supplies. What seems to be of greater concern are the impacts on land use and local public services such as education in a region that is currently very sparsely populated and not provided with excess capacity in service facilities.

In the next chapter we briefly survey the prospects and problems of developing this resource, and then describe the model we shall use to measure the economic, demographic, and fiscal impacts of a number of alternative strategies for developing it in one area of eastern Montana.

[16] National Academy of Sciences, *Rehabilitation Potential of Western Coal Lands,* a report to the Energy Policy Project of the Ford Foundation.

2 A Model for Forecasting Economic and Fiscal Impacts of Northern Great Plains Coal Development

INTRODUCTION: PROSPECTS AND PROBLEMS

After a quarter of a century of stagnation, the coal industry is expected to experience a dramatic growth following recent large increases in the price of imported petroleum, the decline in yields from oil and gas fields (and limited future prospects in the United States), and the difficulties encountered in deploying nuclear fission technology. The Northern Great Plains Coal Province, until recently only a nominal source of coal, now appears destined to become a significant source of supply for other regions of the country, either by directly exporting coal to steam electric generating stations at load centers along the Mississippi, by converting it into electricity at site for transmission to the Northwest Power Pool, or by converting it to gas.

These energy development activities may involve demands on local areas that are extremely large in relation to the economic base of these areas, and this poses a problem for state and local governments. The influx of additional people in the wake of construction activities required to develop mines, and handling, transporting, and processing facilities in the Eastern Powder River coal basin of northeastern Wyoming, for example, has already had a major impact on the town of Gillette. Demands on local public services have been made with which neither Gillette nor other neighboring communities are able to cope. This has prompted the

13

term "Gillette syndrome" to characterize the malaise attending the effect of the current boom on sparsely populated northeastern Wyoming.

The impact of this development on the natural environment has also been a source of concern. The coal in the Northern Great Plains is characterized by the depth of its seams and its relative nearness to the surface over wide areas of the province. As a consequence, large tracts are sus-ceptible to strip mining. The proposed developments all contemplate large strip-mine operations (10 million tons per year, per mine, being a convenient representative of the output level). The prospect of many such mines in the region suggests the disturbance of tens of thousands of surface acres (there being roughly 90 million acres in this representative physiographic region). Erection of conversion plants, transport facilities, and other related land-demanding activities can have a significant impact on the landscape and the productivity of the land. Erosion and stream sedimentation may attend the mining and construction activities, and it is not established that the deterioration of fish and wildlife habitat resulting from this activity can be fully mitigated, even with land reclamation after the mining activity. However, if the disturbed lands are carefully restored to their original contours, as required by state law in Montana, pilot reclamation projects have demonstrated that the visual attributes of the landscape, at least, can be restored, provided revegetation to stabilize slopes is successful. We will have little more to say about the environmental implications of Northern Great Plains coal development in this report.[1]

A related question, which may be even more important, is whether the region's supply of water will be adequate for the contemplated development.[2] There are basically three sources of water here: precipitation, surface water, and groundwater. Precipitation, in itself, is not much of a factor because rainfall is concentrated in winter and spring, when demand is light, and is quickly carried away to surface streams. In the particular area of eastern Montana on which we focus, there are three major streams, replenished by rain and snow melt, and a number of possible dam and reservoir storage facilities. Groundwater varies from site to site.

The difficulty is that the currently available water during seasonal low flows has already been allocated to other uses, largely irrigation. Yet it is clear that coal development, and in particular the operation of gasi-

---

[1] Impacts on air and water quality are being evaluated in a companion study.
[2] The following observations draw heavily on the report "Environmental Impacts of Fort Union Formation Coal Development," by Constance Boris. Prepared for Resources for the Future, Washington, D.C., 1975.

fication and electricity generating plants, will require large amounts of water. In fact, competing claims to water by ranchers, municipalities, energy developers, and environmentalists have been in the state courts for decades. However, something new has been introduced in the current debate about water for coal: the "federal reserved water rights" doctrine.

This doctrine, which actually dates from the 1880s, and was confirmed in a U.S. Supreme Court decision in 1908, states that when the United States reserved its vast tracts of natural forests and parks, it retained the implicit right to take whatever water is "reasonably" necessary to manage the tracts. Under it, for example, the Forest Service could conceivably divert water from irrigation to energy, by insisting on substantial minimum flows in streams through national forest lands. Ostensibly, this would be used to maintain the well-being of the forests, but would also result in substantial supplies available to downstream energy users. Of course, interest group politics may be expected to intrude on judicial application of federal reserved water rights. A coalition of ranchers and environmentalists could perhaps delay or even prevent substantial diversion to coal operators, much as they have the sale of currently unused water from Bureau of Reclamation reservoirs.

To sum up, it appears that the demand for Northern Great Plains coal may be quite robust, but the rate of expansion of output in the short run is dampened by lack of facilities—mining equipment, and especially the local public services and facilities needed by the construction workers, miners, their families, and others who arrive as a result of the primary activities. In the long run, the extent of the expansion may be constrained by complementary resource (water) supply stringencies, and by water and air quality standards. Finally, in some areas it is conceivable that the local political process will be dominated by those who are not sympathetic to a level and intensity of development that would threaten traditional life-styles and cultural values, and this will place constraints on the character and intensity of the energy development.

ECONOMIC AND DEMOGRAPHIC IMPACTS: FORECASTING METHODS

We believe that a quantitative evaluation of the economic, demographic, fiscal, and environmental impacts of development would go a long way toward placing the issues raised by the proposed development into some perspective. For example, are the concerns that have been expressed derived from highly unrealistic development scenarios? On the other hand, to what extent are concerns for the adequacy of local fiscal structures, and the natural and human environment, justified, even at the rather modest levels of new development taking place in a basically low-density ranching economy? The consideration of issues giving rise to such concerns is likely to benefit from a successful effort to put quantitative boundaries around anticipated impacts from different intensities and mixes of energy development activities. In this manner it would be possible to consider different options and tradeoffs among them to better inform national, state, and local public decision makers on use of the region's resources.

This study experiments with some methods of analysis that are novel, at least insofar as application to Northern Great Plains coal development is concerned. For this purpose we are selecting for a prototype application the two potentially most significant counties in the Fort Union Formation, Big Horn and Rosebud, in Montana. In this and the following chapters we attempt to determine the economic and demographic impacts of a number of different coal development strategies in this area of eastern Montana. By different strategies we mean different levels, time patterns, and locations of mining activity, and whether or not energy conversion facilities such as steam electric power plants or coal gasification plants are constructed in the vicinity of the mining activity. For the affected counties and the state, and for each strategy, estimates are made of the time patterns of direct and indirect growth in output, investment and employment by sector, population (and its characteristics), and income. In chapters 5–8 we consider the implications that the estimated changes in level and composition of population and economic activity will have for state and local tax revenues. This exercise is also carried out for each of the strategies or energy development scenarios.

In the remainder of this chapter we spell out in some detail the econometric model that will be used to estimate economic and demographic changes. The next chapter describes the different scenarios and traces their impacts.

Before we get to the econometric model, it will be useful to understand why we have adopted it in preference to a number of frequently seen alternatives: projections of past trends, economic base multiplier methods, and regional input-output models. Typically, analyses of the economic future of a region, perhaps in response to some contemplated policy change, such as reduced defense expenditures, employ some variant of one of these methods. We briefly indicate the drawbacks of each, for our purposes. The following discussion is not intended to be an adequate description of the various techniques of regional analysis.[3]

Simple projection, or extrapolation of past trends of such economic variables as output and employment by sector, or of demographic variables such as the school-age population, clearly are not adequate for measuring the impact of a major new development. This is particularly true if, as in our case, the development is quite large relative to the current economic base. In this case we can be fairly certain that past trends will in fact be modified in some way.

Economic base multiplier methods offer some improvement over simple extrapolation. The multiplier methods divide economic activity in a region into two types: basic and nonbasic. Basic activity produces output for export, and other goods and services are nonbasic. Account is taken of the new development by specifying, exogenously, a new level of basic employment. In our case this might mean employment in the new coal mines and energy conversion facilities. Total employment (basic plus nonbasic) and population are then forecast on the basis of multipliers, the ratio of total to basic employment, for employment, and the ratio of population to basic employment for population. The problem, however, is that the multipliers are derived from the current level and composition of employment in the region. For the forecasts to be accurate, the multipliers must remain constant, and there is no reason to expect them to do this in the face of dynamic change in the region's economy. Another problem with this approach is that it is much too aggregative. The basic-nonbasic split, rather arbitrary to begin with, does not capture interrelationships among industrial sectors, or changes in them over time.

This is no problem for the regional interindustry or input-output models, which are explicitly concerned with the disaggregated structure of production: how much of each of a variety of separate inputs are required for an increment to some regional output. Given a knowledge of these technical production relations, it is possible to determine output in each sector consistent with a new bill of final demands and supply of the region's "primary input," labor. There are, however, a number of problems with the regional input-output approach. To begin with, final

---

[3] The interested reader may wish to refer to Walter Isard, *An Introduction to Regional Science,* for a review of regional methods. For a couple of informative reviews of Isard, see Niles Hansen and John Quigley, both in *Land Economics,* August 1976.

demand, though disaggregated, is determined exogenously. Clearly, we would prefer that demands for goods and services in the region be determined endogenously, in response to the proposed new development and the changes in the economy it triggers.

Another drawback of these models is that the input-output coefficients, reflecting the amounts that industries in the region buy from other industries in the region, are fixed. National interindustry models have been criticized for this reason, but the problem is even more serious on a regional level, since movement of firms and industries into or out of the region will almost certainly affect the (assumed fixed) coefficients.[4] The search by firms and owners of resource inputs (including labor) for higher returns in turn ensures that this movement will be a pervasive feature of the region's economic landscape. Ideally, we would like to explicitly model this sort of maximizing behavior.

A final—and probably fatal—disadvantage of the input-output approach for our purposes is that it sheds no light on the dynamic adjustment of the economy to the new equilibrium level and composition of output. But it is this process of adjustment that is crucial in studying the effects of introducing a major development of energy resources into the regional economy of the Northern Great Plains. Probably the heaviest impact, for example, on local public services and water resources will come with the early construction phases of the development, and not with the longer run operation of a set of mines.

This is clearly seen in the figures in chapter 3. Looking, for example, at figure 3-1b, Rosebud County aggregate employment for 1975–90 (in each of the energy development scenarios), we see that by 1985, employment in scenario IV, involving the heaviest development, has stabilized at about 8,700. But it builds rapidly from about 4,500 in 1975 to a peak of about 11,400 in 1982. A model which focuses on the long-run equilibrium and ignores the transition would clearly miss the key development from the planner's point of view—the rapid influx (and departure) of workers in the construction phase.

The foregoing remarks are not intended as an exhaustive or balanced evaluation of methods of regional economic forecasting. Rather, they are meant only to indicate the drawbacks, for our purposes, of some frequently used methods. We feel that a better alternative exists: one that allows for economizing behavior on the part of firms and individuals, and

---

[4] Both problems—exogenous demand and fixed coefficients—also beset interregional input-output models. In addition, interregional models are hampered by a lack of interregional trade data.

describes the resulting changes over time in the region's economy—all of which the methods discussed above fail to do in one respect or another. Our candidate is the multiregional, multiindustry econometric forecasting model developed by Curtis Harris.[5]

THE ECONOMETRIC MODEL

We shall not attempt to simply duplicate Harris's comprehensive description of the model. Instead, we first provide an overview of the way in which the various pieces fit together. That is, we try to understand how developments in one sector, region, or period affect developments in others. This analysis is extended to the energy policy scenarios and their links with other economic activities described in the model.

The next step is to identify a number of key behavioral relationships that drive the model, ignoring the less important ones, the accounting equations, and so on. We also indicate and interpret the major results, that is, the estimates of behavioral parameters for each equation presented. Then in chapter 3, five scenarios for investment in energy development in eastern Montana are described, and their implications for output, employment, income, and population movements are obtained with the aid of the forecasting equations.

An Overview

First, and most important, the model is recursive. Forecasts for a given year (t) are made on the basis of data for the previous year ($t-1$), and so on. This is what allows us to trace the time paths of the economic activities within a region, including their adjustment to such developments as the introduction of coal mines and conversion facilities. The basic regional units of observation are counties—all of the 3,111 counties in the United States. This gives the forecasts an exceedingly fine character, but the smallness of the units can also lead to problems in an application such as ours. We shall have more to say about these problems after presenting further details of the model. There are 99 industry sectors, 4 extra labor sectors (different levels of government) for employment and earnings, 69 equipment purchasing (investment) sectors, corresponding to (combinations of) industry sectors, and 28 types of construction. Population is divided into 4 age groups, and 2 races.

---

[5] Curtis C. Harris, *The Urban Economies, 1985: A Multiregional, Multi-Industry Forecasting Model.* See also Curtis C. Harris and Frank E. Hopkins, *Locational Analysis.*

It is important to note that, though it contains many equations, the model is not simultaneous. Estimation proceeds in a sequential fashion, with all of the independent variables exogenous in each equation. For example, as we indicate just below, the change in output by sector and region is a function of a number of lagged independent variables. This change in output then becomes an explanatory variable in the equation for the change in employment, and so on. Since the independent variables are exogenous in each equation, estimation is by ordinary least squares.

The behavioral relationships are obtained from 1965–66 data, the most recent available at the time of estimation. Most of these data come from the U.S. Bureau of the Census publication, *County Business Patterns,* for 1965 and 1966, and are described in detail in the Harris volume, *Urban Economies, 1985,* cited earlier. Work is now under way to reestimate on the basis of more recent data, and 1970 data are the base for forecasts. It is also possible to introduce still more recent data and projections, as we shall for investment, output, and employment in coal mining and related energy conversion activities.

How does the model actually work? That is, what drives it, and how are the various parts linked together? The key concept is that of *Ricardian rent.* This is illustrated in figure 2-1.

The stepped supply curve for a single commodity, call it X, describes the behavior of producers of X with different cost functions. At price P_1, the lowest cost producer supplies X_1 units of X to the market (we assume for simplicity that each producer supplies some fixed amount). At price P_2, the next lowest cost producer supplies $(X_2 - X_1)$ units, and so on. In this fashion, a supply curve, labeled S on the diagram, is built up. If the curve D represents market demand, the equilibrium price of X, P_x, will be given by the intersection of D and S. At this price (P_j on the diagram), the lower cost producers are receiving *rents,* the difference between market price and their cost of production. For example, the lowest cost producer receives a rent of $(P_j - P_1)$, and so on. These are called Ricardian rents after David Ricardo, who suggested they arise in agriculture because of differences in the quality of land. But we may attribute them more generally to *location.* If one piece of farmland is located in a more fertile region than another, it will receive a rent, à la Ricardo. But the farm—or any other commercial enterprise—could equally well receive a rent because it is located closer to some raw material, or to the market, than its competitors.

The existence of such a rent is of course a signal to the competitors, including potential competitors, to shift location. Theoretically, in the

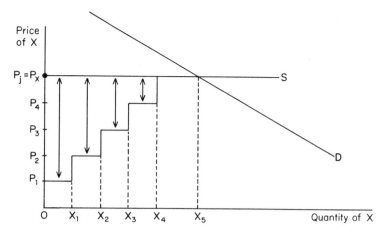

Figure 2-1. Ricardian or location rents.

long run, the quasi-rent will be competed away. This might not be true in some sectors, agriculture for instance, where a particular input cannot be augmented at constant cost, even in the long run. Thus, if the supply of a very fertile grade of land were limited, its owners could continue to receive rents. This is obviously true as well for a particular tract of coal-bearing land. No locational shift would be implied, because there is (we assume) no known additional source of the good quality coal. But in the many related sectors of the economy, i.e., those producing goods and services *linked* to the location-specific agriculture or mining, transitory rents will trigger locational shifts. Firms will move about, not only to get closer to their raw materials, but also to get closer to a pool of trained manpower, to their markets, and so on. This behavior is what drives the Harris model. The long-run equilibrium is never reached, but in every period some firms at least are moving in the direction of reduced costs. The analytical representation of this behavior is given in equation (1), which relates the change in output by sector and region (county) to measures of transport and other costs (such as the local wage rate) and agglomeration variables such as population density and highway congestion. Again, let us emphasize that the existence of location rents for the coal tracts themselves is not at issue here. The opening of coal mines and related energy conversion activities are in any case *exogenous* to the model. That is, we specify a number of possible paths of energy development in Big Horn and Rosebud counties. We then employ the econometric model to calculate the impacts of the exogenous energy activities on other sectors of the economy, on the population, and so on.

At this point, a couple of questions arise. First, why doesn't the economy simply adjust to an equilibrium locational configuration? That is, in the absence of exogenous shocks such as the opening of new coal mines and related conversion facilities, which we shall discuss presently, why should there continue to be movement in period after period? The second question is, what does figures 2-1, which depicts only a single market, tell us about firms confronted with a choice of locating near any of a large number of markets?

Part of the answer to the first question is that, if an industry really is near a locational equilibrium, there will be little movement. This would require both regional input prices *and* demand to be stable over a long period of time, which is not very likely, though certainly possible. Of course, where the configuration of input prices and demand is exogenously shifted, as in our energy scenarios, the adjustment process must begin anew. But in any event, the reason this process takes time is that a firm or industry will typically have a good deal of money sunk in existing plant and equipment. Even if it becomes cheaper for an industry to produce in a new location, all the firms in the industry will not move right away. Some, those who have already depreciated most of their plant and equipment, will. Others will wait and phase out existing operations over a period of perhaps several years. This delayed or partial adjustment is captured by measures of investment and depreciation in equation (1), which explains changes in output by sector and region.

The reasoning here is exactly analogous to that in chapter 1, where we suggested how and when a number of different energy sources would be phased into the economy. We observed that, even if a technological breakthrough made direct solar conversion, or fusion, say, a cheaper source of electricity than existing fuels, the switchover would not take place overnight. There would be a transition period of perhaps several decades as the existing stocks of machines, housing, and other buildings were phased out or converted to the new power source. Now, we are suggesting that this process of gradual depreciation should influence production and location decisions generally.

The second question we raised, about multiple markets, has an easy answer. A situation like that shown in figure 2-1, with inframarginal producers earning location rents, exists in *every* market. The problem, for each producer, is to shift around his production, moving toward (though perhaps never reaching) equilibrium, in which the rent he obtains is the same for each of *his* potential markets. This is described very clearly in *Locational Analysis* (pp. 29–34) and *Urban Economies, 1985* (pp. 33–

37), along with the formal analog, a linear program to minimize transport costs.

How does an exogenous development like the phased opening of a number of new coal mines in a region fit into the process we have been describing? The new development affects both the input prices faced by nearby firms in energy-using sectors and the demand for their output. The effect on input prices is obvious. It is now cheaper for some producers to locate nearer to the coal mines, or to the power plants that may be constructed at the mines. Even if a market had reached a state of equilibrium, the reduction in (some) input prices would upset it, triggering a new round of movements. These supply shifts in turn interact with demand in each market to determine a new pattern of location rents that generates further relocation, and so on.

We noted just above that the energy development would also affect demand. In tracing this effect, we can shed some light not only on it, but on the links between sectors of the model generally. As described in chapter 3, when we specify a development such as the opening of new coal mines in a county, we specify the schedule of dollar outlays for investment in plant and equipment, man-years of construction and operating labor, and tons of coal output. One effect of the outlays is to increase personal income, and hence personal consumption expenditures in the mining county. This in turn leads to an increase in county demand for goods and services in all, or most, sectors. Also, the increase in output within the county—in coal mining, construction, and perhaps other sectors—is translated into a further increase in county demand for goods and services. This is accomplished by means of material input-output coefficients applied to the county output in each sector. For example, suppose output in coal mining increases according to our specifications. According to the input-output coefficients, for each dollar's worth of coal output, there is required so many cents worth of input from, say, iron and steel, rubber and plastics, industrial machinery, and so on.[6] In this way developments in one sector, in one county, are linked to economic activities in other sectors and counties.

All of this can be tied back into the diagrams in figure 2-1. Now we have induced shifts in both supply *and* demand, which affect the location rents of firms and industries. The changes in location rents in turn induce firm and industry movement, thereby further affecting supply and demand

---

[6] These relationships are described by equations (2-32), (2-35), (2-39), (2-40), and (2-46) in Harris, *Urban Economies,* pp. 21–22. The input-output coefficients are from Clopper Almon, *The American Economy to 1975.*

in each market. The process of adjustment in the direction of an equilibrium location pattern continues indefinitely, sometimes in some sectors damping down, other times stimulated anew by developments like those postulated in our energy scenarios.

Some Key Equations

The first major behavioral equation, and by far the most important in the model, relates the change in output in year t, from year $t - 1$, by sector and region, to a number of different input prices, transport costs, measures of investment and capital stock, and agglomeration variables, all in year $t - 1$.[7]

$$\Delta Q_{ij}^{t} = f_{it}(TQ_{ij}^{t-1}, TI_{s_k j}^{t-1}, WR_{ij}^{t-1}, VL_{j}^{t-1}, Q_{ij}^{t-1}, EQ_{hj}^{t-1}, MB_{ikj}^{t-1}, MS_{ikj}^{t-1}, \quad (1)$$
$$CGS_{j}^{t-1}, DEN_{j}^{t-1})$$

$$(i = 1, \ldots, 88)$$
$$(k \leq 1, \ldots, 4)$$
$$(s_k \; \epsilon \max_s q_{si})$$
$$(h \rightarrow i)$$

This looks rather cluttered, but it should be noted that generally only three or four explanatory variables enter significantly in each equation. The superscripts on the variables of course represent time periods, years. The subscript j in all cases indicates a region, that is, a county in this study. Sector is normally represented by the subscript i, in some cases by subscripts h or k. The notation $(h \rightarrow i)$ indicates a matching of elements h to elements i. The variables (with superscripts omitted) are defined as follows. ΔQ_{ij} is the change in output, Q_{ij}, from one year to the next, in sector i and region j. TQ_{ij} is the transport cost of shipping a marginal unit of output from industry i out of region j. $TI_{s_k j}$ is the transport cost of obtaining a marginal unit of input from industry s_k into region j. The subscript s_k refers to the industry, or industries (up to four) believed to be most important in supplying inputs to industry i, as measured by technical input-output coefficients. Transport costs are computed using a linear programming transportation algorithm. Other input prices are WR_{ij} and VL_j. WR_{ij} is annual earnings per worker in labor sector i in region j. VL_j is the value of land per acre in region j. EQ_{hj} is equipment purchases by sector h in region j. This is a measure of gross investment. MB_{ikj} is

---

[7] The change in output variable actually represents the change in the region's share of the nation's output. The national totals come from Clopper Almon's national input-output forecasting model in *The American Economy*.

major buying sector k located in region j that bought goods from sector i. Similarly, MS_{ikj} is major supplying sector k located in region j that sold goods to sector i. These are agglomeration variables, as are CGS_j and DEN_j, measures of highway congestion and population density, respectively, in region j.

This equation is the principal one driving the model. It recognizes that firms seeking profits will move about in response to changing opportunities. Changes in output by sector and region are, as we shall see, negatively related to changes in the prices of various inputs such as land and labor and, importantly, transportation. This sort of economizing behavior is just what some of the alternative regional models discussed earlier fail to reflect.

Note, however, that though firms may move to reduce costs in each period, they do not reach an equilibrium. Instead, they are always gradually adjusting to changes in their economic environment. This is probably a good approximation to the way in which firms actually behave, given the existence of sunk capital as noted earlier.

The results of the estimation of equation (1) are not easy to characterize because different explanatory variables appear to be significant in different sectors. We shall just try to provide an overview of the results, suggesting in a general way how the variables ought to be related in theory, and how they are in fact. Beginning with the transport costs, it is clear that the partial relationship between these and changes in output ought to be negative. All significant transport cost coefficients are indeed negative; in 12 sectors for shipping output, and in 28 for at least one input. Wage rates and land prices should also be negatively related to output changes, and this is confirmed in 6 sectors for wage rates, and 15 for land prices.

Measures of capital and investment, and their effects on output changes are less straightforward. Presumably variation in interest rates would affect industry location, but the money market is national. Variation in construction costs might also be important, but data are unavailable. This leaves lagged equipment purchases, the variable used as a measure of gross investment, and lagged output, the variable used as a measure of depreciation of existing capital stock. The coefficient on equipment purchases is significant in 15 sector equations, and is positive, as one would expect. If lagged output picked up only the effect of depreciation, its coefficient would presumably be negative, where significant. Yet output enters 26 times with a positive sign. The interpretation is that it serves as an agglomeration variable as well, picking up external economies generated by firms in the same industry and located near each other.

For example, firms may draw from the same pool of skilled labor, or may engage in transactions with each other. Output also enters 29 times with a negative sign, and the interpretation is that this represents depreciation.

Two of the other agglomeration variables, major buyers and major suppliers, are believed to act like own industry output, picking up external economies such as, in this case, improved communication among firms.[8] The output of at least one major supplying industry was significant in 18 sector equations, and output of a major buying industry or personal consumption expenditure (if sales made are to final demand) was significant in 34.

The remaining independent variables are the measures of highway congestion and population density. Clearly, the effect of highway congestion should be adverse, and this is confirmed by significant negative relationships between congestion and output change in 25 equations. Population density can cut either way, and this is reflected in the mixed coefficients obtained in 21 significant cases.

Once the change in output is known, a simple accounting relationship gives the output in year t. This in turn determines, along with equipment purchases or investment in $t - 1$, investment in t.

$$EQ_{ij}^t = f_{i2}(Q_{kj}^t, EQ_{ij}^{t-1}) \quad (i = 1, \ldots, 69) \qquad (2)$$
$$(k \rightarrow i)$$

Construction investment in year t is arrived at in much the same way. As theory would suggest, all of the significant coefficients on both output and prior investment are positive.

The change in employment, by sector and region, is a function of output, investment, and the change in output:

$$\Delta EMP_{ij}^t = f_{i3}(Q_{ij}^t, EQ_{kj}^t, \Delta Q_{ij}^t) \quad (i = 1, \ldots, 99) \qquad (3)$$
$$(k = 1, \ldots, 69)$$
$$(k \rightarrow i)$$

The results of the employment equation are interesting, because they are not immediately intuitive. Although the change in employment is positively related in virtually all cases to the change in output, as one would expect, a number of the coefficients on output and investment are negative. The explanation, or at least one possible explanation, for a negative relationship between output and the change in employment, where ob-

---

[8] The term "own industry output" is used to distinguish the effects of outputs of an industry on firms within that industry from the effects of outputs of other industries on these same firms.

served, is that there may be sufficiently strong economies of scale in the industry—and the change in output, remember, is already controlled for. Alternatively, for a given relationship between output and change in employment, the relationship between changes in output and employment is positive, again, as one would expect.

The negative coefficient on investment, which appears in a little less than half the cases, can also be explained in a manner consistent with theory. Specifically, it is possible that where the coefficient is negative, investment is strongly labor-saving. Again, a positive effect on the change in employment is related to change in output.

Population by age, race, and region is determined by births and deaths and also by net migration. The interesting behavioral relationship here is between net migration and the average wage in the region, lagged one period, and a couple of measures of labor surplus in the region, for those of working age.

$$NPM_{arj}^t = f_{ar4}\left(PLS_j^{t-1}, PEC_j^t, \sum_{i=1}^{102} PAY_j^{t-1} \Big/ \sum_{i=1}^{102} EMP_{ij}^{t-1}\right) \begin{array}{l} (a = 2, 3) \\ (r = 1, 2) \end{array} \quad (4)$$

where NPM_{arj}^t is net migration by age group a and race r into region j in year t, PLS_j^{t-1} and PEC_j^t are a couple of rather messy (and in most cases not significant) measures of labor surplus in region j, and

$$\sum_{i=1}^{102} PAY_{ij}^{t-1} \Big/ \sum_{i=1}^{102} EMP_{ij}^{t-1}$$

is the average wage in region j in year $t - 1$.

This equation is the analog for individuals, of equation (1) for firms, and is therefore extremely important. It tells us that individuals, like firms, move about in response to changing opportunities. Presumably, some sort of optimizing calculation underlies the movement. At least, this is what we should say if the relationship between wage and net migration into a region were significantly positive. This is in fact just what emerges from the estimation. Net migration is in all cases significantly positively associated with the lagged wage. Again, we think this is important because it powerfully reinforces the presumption of economic theory, which is questioned by some persons working in the area of human resource allocation. It appears that individuals, at least those of working age, do move in response to relative wages.

The last major behavioral relationship in the model involves the determination of income. Specifically, earnings by sector and region are a function of employment, the prior year's investment, and the prior average manufacturing wage.

$$PAY_{ij}^{t} = f_{i5}\left(EMP_{ij}^{t}, EQ_{kj}^{t-1}, \sum_{h=13}^{74} PAY_{hj}^{t-1} \middle/ \sum_{h=13}^{74} EMP_{hj}^{t-1}\right) \quad (i = 1, \ldots, 99)$$
$$(k \rightarrow i)$$
$$(5)$$

As expected, the relationship between aggregate employment and earnings is positive and significant in all cases. The relationship between the lagged average manufacturing wage [the wage in sectors 13 through 74, as indicated in equation (5)] and earnings in sector i is also positive in virtually all cases. The interpretation is that this represents a "catch-up" effect. Other things being equal, as the average manufacturing wage rises, compensation in sector i will rise. This could be due to the presence of unions, or may be an example of the "Baumol effect," explained by rising productivity in manufacturing and a mobile labor force.[9]

The lagged investment variable appears to represent two conflicting effects, both of which might be expected *a priori*. On the one hand, an increase in the capital stock, holding employment constant, should result in an increase in the productivity of labor, and hence earnings. On the other, it is possible that the new machines call for a lower skill level in the complementary labor input. In this case, earnings would fall. The former effect, which seems more likely, does predominate in more than half of the sectors exhibiting a significant relationship between lagged investment and earnings.

EVALUATION OF THE MODEL

The major drawback of the model, which we hinted at earlier, might be described as a lack of correspondence between the units of observation, which are political (counties), and the units in which economic activity typically takes place, for example, standard metropolitan statistical areas (SMSAs). Suppose the development of new coal mines in a county attracts a number of construction workers and others who sell goods and services to the mining industry there. The model ought to reflect this movement, and it will. But suppose the migrants choose not to live in the county that has the mines, preferring instead some neighboring jurisdic-

[9] William J. Baumol, "Macro-Economics of Unbalanced Growth: The Anatomy of Urban Crisis," *American Economic Review* vol. 57, no. 3 (June 1967) pp. 415–426.

tion. Then, the model will overestimate. This might not be a problem if our study area included all of the counties in an economic area, assuming these areas could be readily determined. In this study, however, just two counties with mining and related energy activities are considered. And as a matter of fact, as discussed subsequently in chapters 7 and 8, there is some population spillover from Big Horn County to nearby Sheridan, Wyoming.

But this may turn out not to be a serious problem as long as we know the direction of the forecast error. In particular, if we know that we have *overestimated* population, say, in Big Horn County, we can be confident that the implied public expenditures are also overestimated. And if we should learn that Big Horn's share of projected coal tax revenues is sufficient to cover the implied expenditures, we would not need to estimate precisely the induced migration.

In other words, in the absence of information that would enable us to be more precise, our research strategy here is to work with inequalities, much as in our earlier volume on the economics of natural environments.[10] There the problem was to assess the costs and benefits of various development projects, where one of the costs was damage to the environment. Although it was not possible to put dollar values on all of the damage, generally we did not have to. Similarly, we probably overestimate expenditures in some instances in this study, but if tax revenues were to exceed the upper bound expenditure levels, the indicated correction would not change our conclusion.

Of course, any regional model—economic base multiplier, input-output, whatever—when applied to such small units as counties is likely to generate a good deal of noise, and perhaps also some systematic error. We admit this is a difficulty with our model, but have also suggested that its usefulness in sophisticated application is not necessarily impaired on that account.

Another possible criticism of the model is that it is not applicable to the area, the two (large) counties in Montana we have chosen to investigate. The model estimates are based on data from all U.S. counties, and the concern would be that the estimated relationships will not forecast accurately for Montana. On the other hand, the fact that neither of the Montana counties looks very much like the "typical" U.S. county at this particular time is not a sufficient reason for rejecting the forecasts for

---

[10] John V. Krutilla and Anthony C. Fisher, *The Economics of Natural Environments*.

these counties. It is important not to confuse different data points with different structural relationships. For example, assume we know that Big Horn County has very little food processing activity relative to other counties. The model does not claim that it has more of this activity than we know it has. It does claim that, should certain preconditions for expansion, such as a fall in transport costs or other input prices, occur, then expansion will follow. Underlying structural relationships are assumed to be the same across counties—profit-seeking firms will respond in much the same way to changing opportunities in Montana as in Missouri—but the counties need not have the same mix of activities at any particular moment.

This may be too glib. Because of nonlinearities, there may be problems with the start-up of activities in counties with a very small industrial base, such as Big Horn and Rosebud in Montana. The model may mistakenly call for certain (new) activities, and fail to predict others. But we can, as Karl Marx might have said, stand this criticism on its head. The activities we are most interested in are those primary ones involving the extraction and conversion of coal. The model will obviously not do well predicting such specific public policy-dependent activities as the opening of mines or the construction of power plants. But these data—the schedule of mine openings, the required construction and engineering employment, the mine output, and so on—are fed into the model, superimposed on an existing data base. Further, the same sort of thing can be done for any other sector, or any region where we have information that is more recent or more accurate than the data base or the forecasts dependent on it. Instead of the forecast steel output, say, in county j and year t, we can use (if we have it) a better, independent estimate of output.

Perhaps, where this sort of information plays an important role in generating model results, the model is better regarded as a calculating engine than a crystal ball and should be used so that it works for the analyst, rather than the opposite. In any event, once the information, for instance, about coal production, is in the model, it can "take advantage of" the relatively sophisticated behavioral relationships there to generate other forecast changes in a region's economy.

CONCLUDING REMARKS

This completes our description and evaluation of the multiregion, multi-industry forecasting model that will be used to determine the economic and demographic impacts of a number of alternative coal development plans in two counties of eastern Montana. These impacts are in turn part of the data used in assessing the fiscal impacts of the various development scenarios in chapters 4–8. In the next chapter, we spell out in detail these energy development futures, and present in highly aggregated form the simulated impacts on population, employment, and income for each of the affected counties. In closing this chapter, we candidly acknowledge the limitations of the model, and the difficulties in applying it as we propose to. As we have suggested, it may be most accurately represented as a computational tool, generating impacts on the basis of detailed data input from outside the model. But even if this is the way we choose to interpret the results, it remains our view that, for all its defects, because of its theoretical advantages and very large data base, the Harris model is probably the best available mechanism for generating the information needed to assess local fiscal impacts.

3 Quantitative Economic and Demographic Impacts of Alternative Coal Development Strategies

INTRODUCTION

The outline of the plans and prospects for development of Northern Great Plains coal is somewhat indistinct. There appears to be genuine local enthusiasm for rapid development in the Eastern Powder River Basin in Wyoming. This development may be temporarily stalled, or the rate slowed, by the somewhat limited legal recourse of the environmental groups opposing development. It seems unlikely, however, that it will be halted permanently. In Montana, unless a split develops in the present alliance between rancher and environmentalist, substantially greater restrictions may attend efforts to exploit the Fort Union Formation coals. In any event, whether or not a different combination of mining, transportation, and conversion activities would arise out of a difference in the local political environment, it may arise out of other causes, for example, resource or environmental constraints. It is desirable then to be able to evaluate different energy development strategies, or choices of different combinations of extraction and conversion options, to determine whether constraints will be binding, and if so, which, and whether an alternative strategy with respect to activity mixes may not better meet the objectives of a local area or a state.

FOUR COAL DEVELOPMENT STRATEGIES

In this study we shall consider four development scenarios for the Big Horn and Rosebud county areas of eastern Montana's Fort Union Formation. The first and simplest scenario represents the "base case" and reflects the present (1975) level of coal development of roughly 20 million tons per year with no energy conversion facilities postulated other than those currently going into operation in the Colstrip area of Rosebud County.[1] Scenario II represents an addition of roughly 22 million tons annual production, to be reached by 1980, and reflects the best estimates of individuals concerned with coal development in Montana based on prospects or plans of the coal industry. Scenario III considers a substantially larger increase in production capacity, consistent with "Project Independence" goals. Specifically, this means four new 10 million ton per year mines in Rosebud County and two new mines of equivalent size in Big Horn County. The first is assumed to begin development in 1975, followed by two more in 1976 and one each in 1977, 1978, and 1979 as indicated in the upper portion of table 3-1. For each mine the data include dollar outlays for investment in plant and equipment, man-years of construction and operating labor, and tons of coal output.[2]

Also, we might note that the information refers to a typical truck and shovel strip-mine operation currently thought to be the appropriate mining method for the Northern Great Plains. This differs somewhat from the kind of operations all of the coal companies had previously anticipated, in which the overburden was to be stripped by dragline. The costs associated with the truck and shovel operation are somewhat higher, but the amount of coal recoverable is greater. Further, the restoration of stripped areas becomes easier with the possibility of segregating the overburden for reuse in rehabilitation, which only exists for the truck and shovel operation.

The six-mine scenario, just described, represents a policy of mining for export only. This is consistent with current thinking in the state of Montana. But we felt it would be useful to assess also the implications of a larger and more balanced complex of energy activities there. Accordingly, we consider a scenario (IV) involving the same schedule of investment in mine capacity, and in addition the construction and operation of two power plants in Rosebud County and a coal gasification plant in

---

[1] Two 350 MW units have been under construction at Colstrip, Montana, and were completed in 1975 and 1976.

[2] These data were obtained from private industrial sources, and are therefore confidential.

Table 3-1. *Sequence of Mine Development, Eastern Montana, 1975–85*

County	1975	1976	1977	1978	1979	1980	1981	1982	1983	1984	1985
Rosebud	M-1	M-2	M-3	M-4	M-5	M-6					
Rosebud		M-1	M-2	M-3	M-4	M-5	M-6				
Big Horn		M-1	M-2	M-3	M-4	M-5	M-6				
Rosebud			M-1	M-2	M-3	M-4	M-5	M-6			
Big Horn				M-1	M-2	M-3	M-4	M-5	M-6		
Rosebud					M-1	M-2	M-3	M-4	M-5	M-6	
					Sequence with coal conversion facilities added						
	M-1	M-2	M-3	M-4	M-5	M-6					
		M-1	M-2	M-3	M-4	M-5	M-6				
		M-1	M-2	M-3	M-4	M-5	M-6				
		P-1 (R)	P-2	P-3	P-4	P-5	P-6				
			M-1	M-2	M-3	M-4	M-5	M-6			
					P-1 (R)	P-2	P-3	P-4	P-5	P-6	
				M-1	M-2	M-3	M-4	M-5	M-6		
					M-1	M-2	M-3	M-4	M-5	M-6	
							G-1 (B)	G-2	G-3	G-4	

M = Mine; P = Power plant; G = Coal gasification facility; (R) Rosebud County; (B) Big Horn County. Number following alphabetic character indicates the year in the mine/facility construction period.

Big Horn County. As indicated in the lower portion of table 3-1, construction on the first power plant is assumed to begin in 1976, on the second in 1979, with completion in 1981 and 1984 respectively, and coming on line in 1982 and 1985. Construction on the gasification plant is assumed to begin in 1981 and end in 1984.

Each of the power plants is of the coal-fired steam electric type, with natural-draft evaporative cooling towers and sulfur dioxide control equipment. Capacity is 2600 MW for each. Our data are dollar outlays on plant and equipment (1974 prices), broken out separately for each year during the development period, and employment during the operating phase.[3] The gasification plant is assumed to have a capacity of 250 million cubic feet of gas per day, and will also account for a certain amount of sulfur and ammonia. There also are data on yearly expenditures for plant and equipment and man-years of construction and engineering labor during the construction phase. Data are also provided on employment during the operating phase.[4]

PROJECTED ECONOMIC AND DEMOGRAPHIC
IMPACTS OF COAL DEVELOPMENT STRATEGIES

Before presenting the projected population, income, and employment resulting from the postulated changes in coal mining and energy conversion, it might be well to recall the principal purpose of this research report and the context within which the analysis proceeds. This study is primarily concerned with assessing the fiscal impacts of development on local jurisdictions. And as suggested in the preceding chapter, although fiscal impacts are ultimately concerned with political jurisdictions, the economic impacts which we are attempting to evaluate in this chapter work themselves out through a different set of processes that, depending on circumstances, may and usually do, transcend local political jurisdictions. If the political and economic areas coincided exactly, then there would be close correlation between economic changes and their fiscal implications. But even in large political jurisdictions such as nations, there is likely to be some measure of imperfect correlation between, say, employment and population changes, owing to short-run changes in the pat-

---

[3] These data are from *Power Plant Capital Cost Trends and Sensitivity to Economic Parameters,* U.S. Atomic Energy Commission, Division of Reactor Research and Development, 1974.

[4] These are from *Report to Project Independence Blueprint,* Federal Energy Administration, prepared by the Inter-Agency Task Force on Synthetic Fuels from Coal, under the direction of the U.S. Department of the Interior, 1974.

tern and composition of international trade, among other things. The relationships between economic changes and their fiscal implications become increasingly erratic as the size of the political jurisdiction and the diversity of its economic base are reduced.

Accordingly, while all the normal difficulties associated with economic forecasting are encountered in this exercise, there are in addition some resulting from the noncorrespondence between the areas in which the economic and political (fiscal) impacts are played out. For example, it is not entirely certain that the employment induced in the service sector, and the population associated with it, will all appear in the county in which the mine or energy conversion activity will take place. Individuals familiar with the Montana scene, for example, have suggested that a significant part of the service sector employment, at least initially, will occur at the traditional regional trade center of Billings, in Yellowstone County, and perhaps to a lesser extent at Miles City in Custer County, rather than Rosebud and Big Horn counties.[5] Second, and perhaps as significant as the failure of all of the ancilliary service employment projected by the model for the county in which the primary energy extraction and conversion takes place, is the possibility that not all of the population associated with the employment in the *primary construction and operation* of new energy facilities will be domiciled in the county of employment. For example, as we note later in chapter 7, the population corresponding to the increment in employment in Big Horn County may be shared with Sheridan County, Wyoming, where initially, at least, and for the scale of development experienced to this point, the services that are available through residence in Sheridan attract families of employees engaged in mine development and operation in Big Horn County.

Recognizing the great difficulty in making exact forecasts, even if we are alert to the areas of likely divergence between model outputs and what will be experienced in the actual situation, we have to choose between a counsel of pragmatism or paralysis. Our philosophical predisposition is toward the former. If we are ingenious enough to have the model work for us, rather than the opposite, there is ample justification for proceeding, on the understanding that model outputs are likely to represent upper bound estimates for employment, income, and population, and hence demand for public services. But as we argued earlier, if the financing of public services is suspected to be a possible constraint on the extraction

---

[5] Personal communications with Professor Paul Polzin, University of Montana, Missoula.

and conversion of the fossil fuels of the Fort Union Formation, the upper bound estimates with which we are compelled to work are in precisely the direction of bias which we should prefer for analytical purposes.

Accordingly, we shall proceed in what follows to describe the model outputs as though all of the employment, income, and associated population, and implicitly the public services (discussed in chapters 7 and 8) associated with these will appear in the two counties under investigation. There will be occasions from time to time to point out, as in the case of state personal income taxes generated from the activity in Big Horn County, discrepancies between model outputs and what would actually occur, and the implications of these for our analysis. But generally speaking, we treat the outputs of the projection model as upper bound estimates consistent with the purpose of our investigation.

Employment

Employment effects are reflected by changes occurring in each of the sectors considered in the forecasting model. In figures 3-1a and 3-1b, we aggregate these effects into a single employment series, displaying the projected number of jobs in each of the four scenarios for Big Horn and Rosebud counties, respectively. The model provides similar information

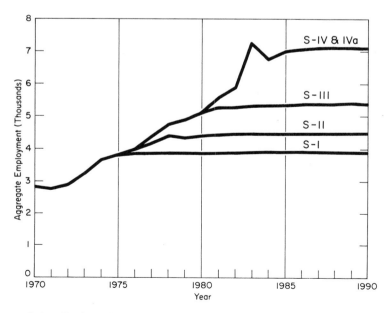

Figure 3-1a. Projected aggregate employment for Big Horn County: 1970–90

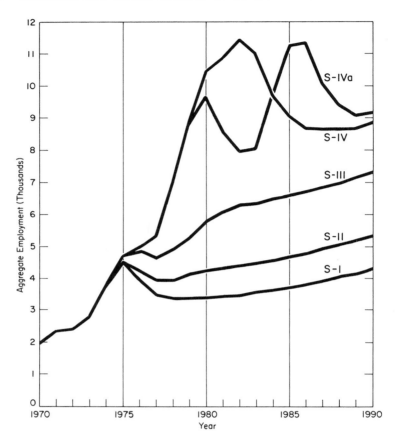

Figure 3-1b.   Projected aggregate employment for Rosebud County: 1970–90

for the state of Montana, but this is not reported here in the interest of brevity.

An examination of the employment effects suggests that without further coal development beyond 1975, coal-related employment will tend to level out. Indeed, for Big Horn County it remains essentially unchanged between 1975 and 1990. In Rosebud County, because of the construction of the steam electric generating units (Colstrip 1 and 2), construction employment declines beginning with the completion of the first unit in 1975 and the second in 1976. But the interaction between the increased coal production and its conversion and the induced secondary effects leads to a moderate growth in employment following the 1975–78 employment readjustment.

Scenario II, which allows for further coal development from 1975 to

1980 of roughly 22 million additional tons annually, induces some further employment. Abstracting from the effects of the construction of Colstrip units 1 and 2, the additional employment obtained from the distribution of the 22 million tons (8.75 million to Big Horn County and 13.87 million to Rosebud) will be roughly 500 persons for Big Horn and 1,000 for Rosebud. The additional 20 million tons per year postulated for Big Horn County under scenario III is attended by another 800 permanent employees. The expansion of production by 40 million tons in Rosebud County corresponding to scenario III results in roughly 1,800 additional employees.

The employment behavior for scenario IV departs significantly from that of coal development for export only. In this case with two additional thermal electric stations postulated for Rosebud County[6] and a gasification plant for Big Horn, we get a significant bulge in employment during the construction phases—almost doubling the employment of the base scenario in Big Horn and more than doubling in Rosebud, where the end of the construction of the first power plant overlaps the beginning of construction of the second. Employment falls off considerably following the peak of construction on these plants. Operating personnel also represent a significantly smaller fraction relative to construction workers in steam electric power generation than in coal gasification. Accordingly, while employment in Big Horn County tends to stabilize after peak construction, with the substitution of operating personnel for construction workers at the gasification plant, the additional employment for scenario IV over scenario III in Rosebud stabilizes at about half the peak level projected during the construction phase.

In order to play out the implications of scheduling the construction of the two 2,600 MW power plants to avoid large construction labor force overlaps from the two plants, a modification (a) of scenario IV is also given in figure 3-1b. Indeed, it would be possible to schedule power plant construction even more uniformly, with care taken to prevent the overlapping of construction among the four 1,300 MW units comprising the two plants. This is not shown, the rescheduling of the two 2,600 MW power stations being used only to illustrate the potential effects of time-phasing construction.

---

[6] This expansion represents two units of 2,600 MW each as discussed in the section on coal development strategies above, and is in addition to the Colstrip units 1 and 2, and not the same as the proposed Montana Power Company's units 3 and 4, which are 750 MW units and are not incorporated into the present analysis.

Population

Since our strategy is to assume that all of the population associated with the increment in employment arising out of direct and induced employment effects of coal development in Big Horn and Rosebud is domiciled in the county in which the employment is attributed, we would expect a rather close correlation between employment and population. The pattern of population increases indeed corresponds rather directly to the pattern of employment for each of the four scenarios (figures 3-2a and 3-2b). Taking scenario I as the base case, we observe a pattern of population stability or slight decline for Big Horn County after 1975 which is

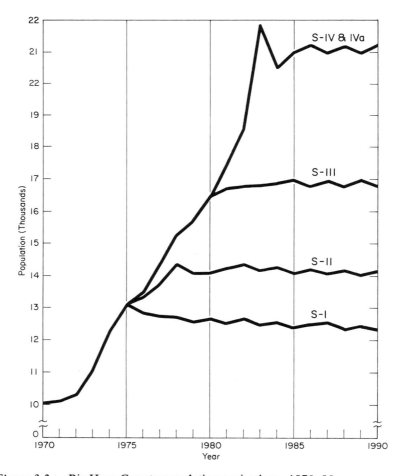

Figure 3-2a. Big Horn County population projections: 1970–90

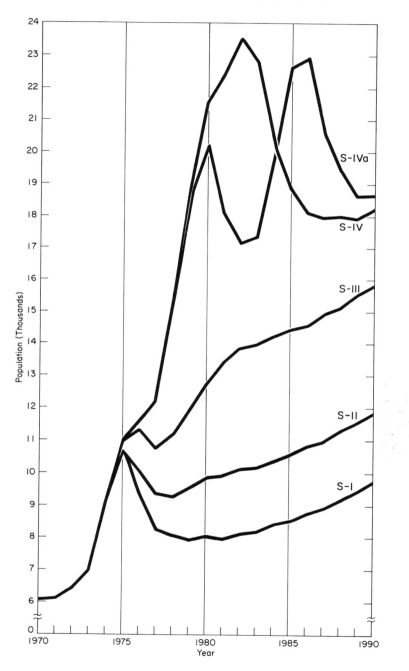

Figure 3-2b. Rosebud County population projections: 1970–90

consistent with the behavior of its population during much of the past three decades. The addition of 700 MW of power at Colstrip and related coal development shows population effects corresponding to the growth in employment after 1979–81 in Rosebud County. The addition of 22 million tons of coal production in scenario II over the base results produces an increase in population of approximately a couple of thousand in each of the two counties. Virtually all of this occurs at or near the time of the expansion in coal production from the roughly 20 to 42 million tons per year between 1975 and 1980.

Scenario III postulates a 40-million ton per year expansion of production in Rosebud County and an additional 20 million tons in Big Horn. This difference in development is reflected in a somewhat larger growth in population in Rosebud County (approximately 4,000) than in Big Horn (approximately 2,500).

Once the effects of the expansion have been experienced, the population tends to stabilize at the level reached shortly after the completion of expanded coal production in Big Horn, but tends to grow in Rosebud, consistent with the projected growth in employment there.

As in the case of employment, the population for scenario IV fluctuates substantially in Rosebud County between the peak of construction and the subsequent operating phase of the energy conversion facilities. This does not appear to be so for Big Horn County, which is consistent with the projected employment in this respect.

Again we draw attention to the fact that the projected population corresponds to the employment levels associated with the postulated development. It is possible, as mentioned above, that while employment will be at the site of coal development, the workers will be domiciled in another county; indeed, this is the case of adjoining Big Horn County and Sheridan County, Wyoming, in another state. This is worthy of note at this point, but more detailed treatment will be deferred until the implications of the dissociation between employment and place of residence for state and local fiscal operations are considered in chapter 7 and subsequent sections of the study.

It should be noted that scenario IV peak population during the construction phase in Rosebud County will be roughly two to three times the peak population of the base case, and nearly double that associated with scenario III. In Big Horn County the difference between scenario IV peak construction population and that of the base case for the same year is about 70 percent. The difference between the peak population of scenarios

III and IV, however, is only about 30 percent, and in this case there is little change between peak population during the construction phase and the stabilized population in the subsequent operating phase of expanded energy development.

Again we observe that as in the case of employment, the peak in the case of Rosebud County could be increased or dampened by the assumption made regarding the *scheduling* of the introduction of new steam electric generating capacity. The construction forces for the two plants need not overlap—indeed, since the two power plants are made up of two 1,300 MW units each, these could be developed in sequence, reducing the peak construction employment by a factor of 2 or 3. Experiments with different scheduling to minimize the impact of the postulated 5,200 MW expansion in power generating capacity could be run with the forecasting model (IVa being an example), and of course, would be rather standard operating procedure were the model to be used for planning coal and energy development in the area. No doubt a schedule could be set that would result in a greatly moderated ratio of peak construction to stable operating labor, and hence fluctuation in population. The present schedule is only one of any number that can readily be postulated and evaluated, given the model's capability.

Personal and Per Capita Income

Although employment and population may stabilize after the first impact of coal development works its way through the system, personal income tends to increase for each of the scenarios in both counties, including the base case (figures 3-3a and 3-3b). This is due to the increase in productivity incorporated into the forecasting model, despite the fact that incomes are reckoned in dollars of constant purchasing power.

Apart from the mild uptrend, the pattern of personal income projected for the four scenarios follows the pattern of employment and not a great deal more needs to be said about it here. There is, however, an interesting difference in the level and rate of growth of per capita income between Big Horn and Rosebud counties (figures 3-4a and 3-4b). It appears to be because the amount of new employment relative to the base case is greater for Rosebud than for Big Horn, and this higher paying employment in mining and related activities is reflected in an increasing per capita differential in scenarios III and IV.

In summary, our model is capable of taking any postulated coal and related energy resource activities and simulating the economic and demographic impacts by county, or by county and larger regional aggregations. These impacts in turn provide the data for further economic analysis, in particular of fiscal impacts. In the next chapter we begin the study of fiscal impacts, looking first at the structure of state tax laws and even federal legislation where royalty-sharing provisions of mineral leasing are relevant to our study.

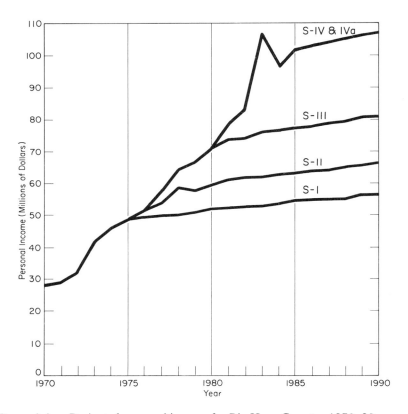

Figure 3-3a. Projected personal income for Big Horn County: 1970–90

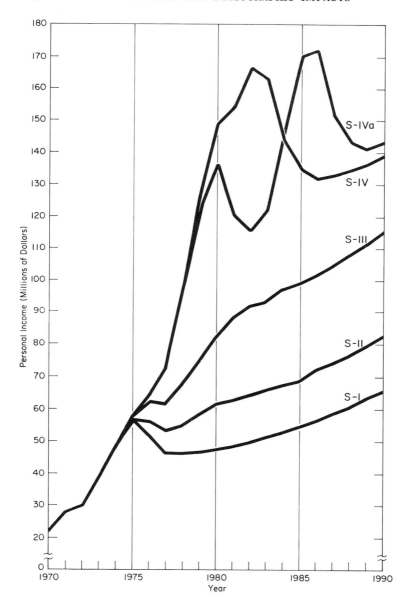

Figure 3-3b. Projected personal income for Rosebud County: 1970–90

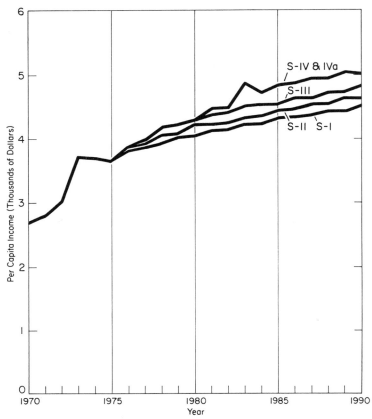

Figure 3-4a. Projected per capita income for Big Horn County: 1970–90

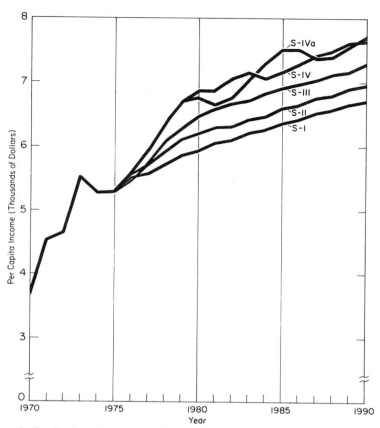

Figure 3-4b. Projected per capita income for Rosebud County: 1970–90

4 Modeling the Montana Tax Legislation

INTRODUCTION

Coal development in eastern Montana will result in increased employment, personal income, and population in the counties in which it takes place, as we observed in chapter 3. These employment and income effects represent transactions predominantly in the private sector. However, there are also economic implications for the public sector that involve changes in the flow of public revenues and, in turn, related demands for public services. In this chapter we review the sources of public funds that are associated with coal development. In subsequent chapters we estimate the various tax yields by jurisdiction and compare them with the implied obligation for expenditures on community services.

This comparison of projected revenues and expenditures may appear fairly naive. Surely the affected communities will be able to exercise some control over tax yields, for example, by varying rates. And public expenditures can be influenced by varying such items as the per pupil allocation of educational resources. An economist with his eye on the long run might then conclude that the local fiscal impact of any change in economic activity resulting from coal development in Montana will be insignificant, since the affected communities can make it so by simply adjusting their tax and spending policies.

A still more sophisticated approach might assume that the communities will adjust their fiscal policy variables in order to maximize some

48

desirable magnitude, such as local property values, or net revenues. If either of these is correct, why bother projecting revenues and expenditures on the basis of current tax laws and customary or mandated spending levels, as we propose to do?

We feel there are at least a couple of reasons for taking this approach. In the first place, even assuming no barriers of any sort to adjustment of rates, we need to know what the revenues and expenditures will be on the basis of *current* rates to know whether, and how, they ought to be adjusted.[1] Our projections are a first step in determining tax and expenditure policy. In the second place, it often is not a simple matter for a local jurisdiction to raise property tax rates, for example, or to institute a sales tax —at least not until it has experienced difficulty borrowing to meet a deficit, or has had to make painful cuts in public services. Thus, there may be a significant lag in the adjustment process.

There are two basic sources of funds in the public sector that result from increased coal extraction and related activities. One source is related directly to the amount and/or value of the coal that is mined. Severance taxes and royalties are examples of this basic source of public revenue associated with coal mining. The other source relates to the increase in incomes and wealth of the community as result of increased economic activity brought about by coal development. For example, the increased direct and indirect employment and income associated with coal development will yield increased state personal income tax receipts. These, of course, accrue to the state's revenue accounts to be used for various state government purposes. But local taxing jurisdictions, the county, town, and school districts, will have sources of increased revenue as the value of real estate increases in response to increased demands for building sites and housing, and as commercial and industrial property becomes incorporated into the local property tax base.[2] In this chapter, then, we review the elements of the Montana tax structure that are related to coal development as a basis for estimating yields associated with different policy options or development strategies.

---

[1] In practice, we do take account of possible variations in two ways. We employ sensitivity analysis, and we examine effects of changes in variables such as scale and density on millage rates in a statistical analysis of cross-sectional data.

[2] The increase in taxable property also implies an increase in the local unit of government's obligation to supply public services that are associated with this change in revenue flows. The interesting question is whether the increased revenue accruing to a taxing jurisdiction is equal to the implied public expenditure obligation within this jurisdiction, and when not, whether there are adequate transfer mechanisms to permit shifting benefits and burdens to achieve more equitable distribution of each through various "equalization" mechanisms.

MONTANA COAL TAX LEGISLATION

The Montana state legislature, meeting during the early part of 1975, overhauled the coal tax legislation that had been in effect since 1973 by enacting Chapter 525, Laws of 1975, Codified at Title 84, Chapter 13, Revised Codes of Montana. The main provisions of the legislation presented as "An Act Revising the Taxation of Coal Production" were to change the severance tax from an absolute sum per ton, depending on its British thermal unit (Btu) content, to a *percentage of its value*, and to change the so-called "net proceeds tax," which was exceedingly difficult to administer, into a "gross proceeds tax." Since this legislation is quite significant, some discussion of specific provisions is warranted.

The 1975 legislation replaces previous legislation that provided for a 12, 22, 34, or 40 cent per ton severance tax respectively, depending on whether the Btu content per pound was 7,000 or less, 7001–8,000, 8,001–9,000, or 9,001 and over. The new law requires that for the 7,000 Btu and lower coals, the tax be 12 cents per ton or 20 percent of the value, whichever is the greater. For coal with a Btu content of 7,001 or higher, the schedule of the previous legislation holds with the added proviso that 30 percent of the value be taxed, using the method that yields the larger amount per ton. In short, for the strip-mined coal with which we shall be dealing, the effective severance tax will be 30 percent of the "contract sales price."[3]

The legal meaning of contract price requires some discussion. The f.o.b. price at the mine is likely to be interpreted in an economic sense as the sales price. But, since a number of different taxes will apply to the coal mined, for legal purposes the "contract sales price" is intended to be a base price, with each of the several taxes on the *value* of coal applied to such a base "contract sales price" in order that taxes will not be paid on taxes that otherwise are incorporated in the f.o.b. contract price. An example of the method by which this tax is computed will be given following discussion of a second important feature of the revised legislation.

Title 84, Chapter 13 also addresses the taxable property value of a mine for local tax revenue purposes that had previously been dealt with under the rubric, "Net Proceeds Tax" (Revised Codes of Montana, 1947, Section 84-5402). Because of the difficulty of ascertaining "net proceeds" and problems with its administration, the 44th Legislature provided for a

---

[3] The legislation provides that any mine owner can deduct up to 5,000 tons of coal mined per quarter before the tax becomes effective. Second, underground mining involves a lesser tax, with 4 percent of the contract sales price being a maximum.

change in procedure involving a gross proceeds basis for taxation. It should be noted that while reference is being made to "net proceeds" and "gross proceeds" taxes, in reality the terms simply refer to a local severance tax in lieu of a property tax determined by assessing the real value of an operating mine. The method of determining the taxable property value of a mine, under the new legislation employing the gross proceeds method, takes the base, or legal "contract sales price" of each ton of coal from a given mine, against which a 45 percent assessment is made. This serves to establish its "taxable property value" under the legislation.[4]

Let us consider then how the new tax legislation is applied to any given ton of coal. If we have the "base price" coal operators negotiate with their buyers at $4 per ton, we have two separate taxes applying as covered by this legislation. If we assume that the levy of a county in which the coal mine producing this coal is located runs in total 200 mills, then the gross proceeds tax would be 0.45 × 0.20 or 9 percent of $4, for a total of $0.36 per ton. This tax would be factored into the f.o.b. price at the mine, along with other "add-ons," but would not figure into the "contract sales price" on which the severance tax would be computed. The severance tax, which at 30 percent yields more than the specific absolute values (retained provisions of the old legislation), would be $1.20 per ton. Both taxes, factored into the price f.o.b. mine would raise the price to the seller to $5.56 per ton, but the legal "contract sales price" for computing the severance tax would remain at the negotiated "base price" exclusive of the permissible "add-ons."[5]

The revenues produced by the gross proceeds method of determining mine property taxes are retained for schools[6] and other county purposes,

[4] Under the new legislation, the "annual gross proceeds of coal mines using the strip mining method" represent class nine "properties" for tax purposes. Class nine property now bears a 45 percent assessment, against which the tax levies of a particular taxing jurisdiction would be applied. For example, a mine producing coal with a base price of, say, $4 per ton, and 120,000 tons, would have a taxable property value of $4 × 120,000 × 0.45, or $216,000. The $216,000 of taxable property, then, would yield tax revenues of varying amounts depending on the millage levies of the taxing jurisdiction. A millage rate of 200 would represent a 20 percent yield on $216,000, or a tax revenue of $43,200 per year for the assumed level of extraction and price of coal. Another way to view the matter is to observe that the 0.45 × 0.20 yields a rate of 9 percent of the gross value of the output.

[5] Admittedly, the language of the legislation might give rise to another interpretation regarding the method for calculating the severance tax, working backward from the f.o.b. price at the mine. The legislative history, however, supports the method suggested above, according to Roger Tippy, legislative counsel and participant in drafting the provisions of the legislation in question.

[6] This is only partially correct. For very large taxable properties, as would be the case of those associated with coal development, a mandated 40-mill minimum required of every school district to support the so-called foundation program may,

with the exception of a 6-mill levy on all real and personal property that goes to the state. The severance tax, however, goes to the state, with specific provisions earmarking the revenues as follows:

	1975–1979 (%)	1980 and thereafter (%)
School equalization	10	10
Local impact and education trust[7]	27½	35
Coal area highway improvement	10	——
County where coal is mined	4	3½
Counties' land planning	1	——
Alternative energy research	2½	4
Renewable resource development	2½	2½
Parks acquisition	2½	5
General fund	40	40

Apart from the *severance tax* and *gross proceeds tax* on coal described above, there is yet another tax based on the value of the mine output. This is the *resource indemnity trust account tax* provided in legislation passed by the 43rd Legislature (Revised Codes of Montana, 1947, Title 84, Chapter 70) in 1973. This legislation provided for an environmental reclamation account in the trust and legacy fund, to be financed by a tax on the gross value of the output of "nonrenewable resource extracting industries." The tax is computed with a constant $25 plus 0.5 percent of gross value of product in excess of $5,000 per annum, extracted by any such industrial firm.

THE MINERAL LEASING ACT

While not part of the Montana tax legislation, the Mineral Leasing Act of 1920, and its 1976 amendment, which is referred to as the Federal

---

indeed is likely to, produce more revenue than needed throughout the entire county to meet a minimum foundation program. The excess of receipts over the aggregate foundation program costs reckoned for all of the school districts in the county in question goes to the state equalization fund for school districts throughout the state which have insufficient property values to support unassisted a foundation program with a 40-mill levy.

    [7] The distribution of the severance tax receipts between local impact mitigation and the educational trust account is worth comment. The 27½ percent earmarked for the two purposes combined is to be divided in such a way that no more than 17½ percent will go for local impact mitigation and less than 20 percent into the educational trust. To simplify computations in the next two chapters where we estimate receipts by jurisdiction, we assume that the maximum allocation to local impact mitigation, and minimum to the educational trust represents an estimation of the distribution of these funds that is good to a first approximation.

Coal Leasing Amendment Act of 1975,[8] are relevant to any discussion of revenues available to the state as a result of coal mining. Taking the 1920 legislation, which is applicable to any coal being mined on the federal lands under leases issued prior to the 1976 legislation, Section 35 of the original Act provides for a royalty of 5 percent of the value of the coal, 37.5 percent of which is returnable to the state where the coal is mined. Under the legislation, such funds are required to be used for road and school purposes within the state, but not necessarily within the county of origin. In any event, they represent funds that are available for dealing with given local impacts at the scene of the mining, at the discretion of the state. The 1976 amendment has a couple of significant changes. Under the new legislation, the royalty has been increased from 5 percent to not less than 12.5 percent of the value of the coal for surface-mined coal. Second, the provision relating to the share of the royalties going to the states has been increased from 37.5 percent to 50 percent.

Two points should be made in passing. One is that whereas the severance and gross proceeds taxes apply to all of the coal that is mined (beyond the small deductible amount), the royalty payments are made additionally only on the coal that is taken from federal lands. In the Big Horn and Rosebud county areas, coal extraction takes place on private, state, and Indian lands in addition to federal lands. Indeed, of the roughly 20 million tons of coal currently being mined, only about a quarter involves federal leases.[9] The second point is that the royalty payments are relatively modest compared to the severance tax. Under the terms of the old legislation, the royalties averaged about 17.5 cents per ton on coal leased and being mined under scenarios I and II, with the state's share roughly 6.5 cents per ton of coal. With both the value of the coal and the percentage of the value increased under the amended legislation on the one hand, and the percent of the resulting royalty that is being returned to the states increased, it is likely that the royalty payments to the state on new leases at higher values will be several times as large—31 cents per ton as a rough approximation. Even so, since the coal taken off the federal lands will be only a fraction of that mined in the state, and the royalty less than half the rate *ad valorem* of the severance tax, with only half the

---

[8] Public Law 94–377, 94th Cong., S. 391 August 4, 1976.

[9] As private and public lands are intermingled, especially railroad properties and government lands, the proportion of the coal coming off the federal lands may vary from month to month. But on average and for the three mines involved (the Westmoreland mine involves Crow Indian lands almost exclusively) the current estimate is about a quarter (personal communication with Douglas Hileman, USGS, Billings, Montana).

resulting yield being returned to the state, the royalty payments on coal taken from the federal lands will continue to remain the smaller portion of the yields to local jurisdictions from revenues accruing directly from the mining of coal.

OTHER RELEVANT MONTANA TAXES

The tax on industrial property may be relevant to coal development even though it is not levied directly on coal. This is the tax that would apply to above-ground improvements at mines, and to facilities for converting coal into other energy commodities such as synthetic fuels and electricity. The standard method of calculation is to take 75 percent of the actual investment costs as the cash value of the property, to which is applied 40 percent to yield the "assessed value." The taxable value is typically 30 percent of the assessed value, or a total of 9 percent of the investment in plant and facilities (i.e., $0.75 \times 0.4 \times 0.3 = 0.09$). Against this "taxable value" would be applied the various levies for school, town, county, and state purposes.

While the taxable value is typically 30 percent of assessed value, provision is made to permit new industrial property to have only 7 percent of its assessed value enter the tax rolls for a period of three years, after which the 30 percent provision becomes effective. It is not clear what, if any, coal-related development would qualify as new industrial property. It is believed that the intent of the legislation is to provide incentives for new, independent enterprises, making it doubtful that an additional facility of an existing coal or related energy commodity producer or converter would qualify under provisions of this legislation (Revised Codes of Montana, 1947, Section 84-301, Subsection 1999, Class 7(a). Supplement, p. 7).

A further consideration relates to the large investment in power-driven vehicles involved in a truck and shovel strip-mining operation and whether these specialized pieces of equipment would qualify as "class two" property for taxable purposes. Vehicles of various kinds, including "motor trucks," enter the tax rolls as class two property at only 20 percent of assessed value. To the extent that a substantial portion of the investment in mining equipment would qualify as class two property, the tax yield level may be affected by 15 to 20 percent.

Finally, one other qualification should be mentioned in connection with the percent of assessed value that enters the tax rolls. For enterprises that are involved in converting primary commodities, energy commodities in particular, the investment in air pollution abatement facilities is allowed to enter the tax rolls at 7 percent of assessed value.

The appropriate percentage to apply to any particular industrial plant's assessed value is somewhat indeterminate, depending on the circumstances and perhaps also rulings by the Department of Revenue, and in disputed cases, the tax courts. Without advance knowledge of the way in which such rulings might go, we will play out the implications of a number of them to test the sensitivity of the tax yield to the nature of the ruling on taxable value.

The corporation license and the *electrical producers' license* taxes may also generate revenues related to eastern Montana coal development. Every corporation doing business in the state of Montana is subject to the corporation license tax. The net income obtained from business conducted in Montana is taxed at 6.75 percent (Revised Codes of Montana, 1947, Title 84, Chapter 15). Since the tax is applicable only to the income obtained from business transacted entirely within the state, very little yield is forthcoming from the operation of firms engaging in interstate commerce. Indeed, given the preponderance of such activity in the area with which we are concerned, the corporation license tax is likely to be of only negligible significance, yielding revenues perhaps not detectable within the range of error of our estimates.

The electrical producers' license tax may be of a different order of significance, however. The completion of two units of the Montana Power and Light's Colstrip plant during 1975–76 suggests increased output of coal-based electrical energy. Moreover, even with a policy of restricting coal conversion to satisfy only domestic (Montana) demand for energy, the announced policy of the Canadian government to phase out deliveries of natural gas and petroleum to the United States is likely to affect Montana severely. A 10-billion cubic feet per year reduction in deliveries was put into effect in the spring of 1975, and the remainder of the 50 billion cubic feet of gas per year imported by Montana is expected to fare no better. It is likely that replacement energy for Canadian natural gas will be electricity produced from eastern Montana coal. The total gross value of electricity generation, or the sales value of the energy produced, is taxable at 1.688 percent. The yields from the tax go to the state's general fund, except for the tax on 0.25 percent of the gross value, which funds the office of the consumer counsel.[10]

---

[10] This is a constitutional office established to intervene on behalf of consumers in public utility rate-setting procedures.

SOURCES OF REVENUE INDIRECTLY RELATED TO COAL

The taxes discussed above are either levied directly on the value or volume of coal extracted, or on the value of investment in plant, equipment, and related facilities involving coal extraction or conversion. There are two additional sources of revenue that are related only indirectly to eastern Montana coal development. One involves the claims the state has on a portion of the increase in personal income related to the increase in employment induced by the coal development. The Montana state personal income tax is a graduated tax varying from 2 percent to 11 percent of taxable income, with an additional 10 percent surtax. Estimated average yields per $1,000 income interval per taxpayer have been prepared by the Division of Research, Montana State Department of Revenue, and are available for use in estimating the increment in revenues derived through the state personal income tax that would be associated with any postulated scenario of coal development.

The other indirect tax is a counterpart revenue for the local taxing jurisdiction. It is based on the increment in the real and personal taxable property that would attend the increases in population associated with each coal development scenario. While this may be the sole source of added revenue to meet added public service obligations for some of the local units of government,[11] our basis for estimating the increase in real and personal property taxes is the least satisfactory. There are six or seven potential classes of real and personal property, with subclasses within some, each of which may be taxed at a different percent of its "true and full value." Perhaps determining the expected composition of assets possessed by new migrants into eastern Montana would be a simpler undertaking than determining the means by which tax assessors estimate the "true and full value" of different items among the total taxable assets of individuals. We shall discuss this problem at the appropriate point in chapter 6 where we address the method of estimating the per capita taxable real and personal property base for migrants attracted to the area by developments in the coal fields and related activities.

---

[11] Since the sole source of tax revenues at the local level is taxable real and personal property, and most school districts and virtually all of the towns are not likely to have mines and taxable industrial property actively operating within their tax jurisdiction, the property tax yield from what is often a rather meager base is the sole tax revenue source. Where any such taxing jurisdictions suffer severe adverse effects from coal-related activities outside their tax jurisdiction, some assistance would likely be forthcoming from revenues collected through the severance taxes earmarked for local impact mitigation in the county from which they are collected.

In the appendix to this chapter, we present the formal computational models that are used in the remainder of the revenue sections of this study to estimate the yield from each of the sources of revenue that is associated with coal extraction, conversion, and related activities.

APPENDIX A

REVENUE MODELS OF MONTANA'S TAX STRUCTURE WITH FUNCTIONAL DISTRIBUTIONS AMONG COUNTY AND SCHOOL DISTRICT EXPENDITURE CATEGORIES

by John V. Krutilla and V. Kerry Smith

1. Severance Tax (ST)

$$ST_{kt} = \sum_{i=1}^{R} \left[\delta \sum_{j \in \Omega_{ki}} P_{jt} Q_{jt} - W_k \Delta_i \delta P_i^* D \right]$$

where:
- R = number of firms mining coal
- δ = severance tax rate
- Ω_{ki} = the set of mines owned by firm i located in county k
- P_{jt} = price per ton of coal for mine j in period t
- Q_{jt} = quantity in tons of coal from mine j in period t
- D = allowed deduction in tons
- P_i^* = maximum P_{jt} for all mines owned by a given firm
- W_k = county k's required fraction of severance tax deduction
- Δ_i = 1 if the firm operates in the county, 0 otherwise

This equation estimates the severance tax revenues for each county as a fixed fraction (δ) of the gross receipts from all mines in the county less the allowed deductions by firm.

The total receipts by the state are obtained by summing the county receipts over the counties.

$$ST_t = \sum_{i=1}^{R} \sum_{k=1}^{S} \left[\delta \sum_{j \in \Omega_{ki}} P_{jt} Q_{jt} - \delta P_i^* D \right]$$

S = number of counties

If we wish to estimate the severance tax receipts returning to the county, then a fixed allocation scheme is applied to severance tax receipts generated in each county.

$$AST_{lkt} = \theta_{lt} \sum_{i=1}^{R} \left[\delta \sum_{j \in \Omega_{ki}} P_{jt} Q_{jt} - \Delta_i P_i^* D \right]$$

where: AST_{lkt} = allocated severance tax receipts to category l for county k in period t

θ_{lt} = fraction of severance tax receipts allocated to expenditure category l in period t

The total of the severance tax receipts returning to the county is simply the sum of the AST_{lkt} over the relevant expenditure categories as:

$$AST_{kt} = \sum_{l \in \psi_t} \theta_{lt} \sum_{i=1}^{R} \left[\delta \sum_{j \in \Omega_{ki}} P_{jt} Q_{jt} - \Delta_i \delta P_i^* D \right]$$

ψ_t = the set of expenditure categories defined in period t to return to the county

2. Mine Property Tax (MPT)—Gross Proceeds Method

$$MPT_{kt} = T_k^g r \sum_{i=1}^{R} \sum_{j \in \Omega_{ki}} P_{jt} Q_{jt}$$

where: r = percentage of gross receipts which functions as taxable value (assumed equal to 45 percent)

T_k^g = county levy on taxable mine property for variant g. Three variants corresponding to 0.027, 0.031, and 0.036 of the taxable value in dollars

R = the number of firms

P_{jt} = price per ton of coal for mine j in period t

Q_{jt} = quantity in tons of coal from mine j in period t

Ω_{ki} = the set of mines owned by firm i located in county k

This equation designates the proceeds of the mine property tax for each county as a given percentage of the taxable value which is defined to be 45 percent of gross receipts of the mines in the county. To calculate the proceeds of the state tax on mine properties, the taxable values are summed across counties and a separate tax levy is applied as given below.

$$MPT_t = T^{st} r \sum_{k=1}^{S} \sum_{i=1}^{R} \sum_{j \in \Omega_{ki}} P_{jt} Q_{jt}$$

where: S = number of counties in the state

T^{st} = state levy (assumed equal to 0.006 of the taxable value in dollars)

In order to allocate the county tax receipts to functional areas, a fixed percentage is applied to each county's tax receipts as given below.

$$AMPT^q_{kst} = d_{ks}T^q_k r \sum_{i=1}^{R} \sum_{j\epsilon\Omega_{ki}} P_{jt}Q_{jt}$$

where: $AMPT^q_{kst}$ = the amount of MPT_k allocated to category s for county k in time t with variant g

d_{ks} = distribution weight for county k in service category s

3. Resource Indemnity Trust Account Tax (ITAT)

$$ITAT_{kt} = \alpha \sum_{i=1}^{R} \sum_{j\epsilon\Omega_{ki}} [P_{jt}Q_{jt}] \qquad \text{for all} \qquad \sum_{j\epsilon\Omega_{ki}} [P_{jt}Q_{jt}] > \$5,000$$

where: α = tax rate (specified as 0.5 percent)

R = number of firms engaged in mining operations in the state

Ω_{ki} = the set of mines owned by firm i located in county k

P_{jt} = price per ton of coal for mine j in period t

Q_{jt} = quantity in tons of coal from mine j in period t

The indemnity trust account tax revenues for the county are specified to be a fixed fraction of the gross receipts of all firms. These gross receipts are estimated as the price times quantity for each mine, by firm and county.

In order to derive estimates of the annual state receipts, these revenues must be summed across counties as given below

$$ITAT_t = \sum_{k=1}^{S} ITAT_{kt} = \alpha \sum_{k=1}^{S} \sum_{i=1}^{R} \sum_{j\epsilon\Omega_{ki}} [P_{jt}Q_{jt}]$$

where: S = number of counties.

4. State's Share of Royalties (SSR)

$$SSR_t = \eta \sum_{j\epsilon B} \bar{r}_j Q_{jt}$$

where: η = state's share of the royalties (assumed to be 0.375, and 0.5 on leases after the 1976 amendment to the Mineral Leasing Act of 1920)

$\bar{r}_j$ = royalty rate for mine j

Q_{jt} = quantity in tons of coal from mine j in period t

B = the set of mines on federal lands

This equation calculates for each mine on federal lands the royalties resulting from coal production in each period. Then the state's share is calculated as a fixed fraction of the royalty.

The present value of these royalties may be calculated by summing over t and discounting

$$PV\text{-}SSR = \sum_{t=1}^{T} \left[\eta \sum_{j \in B} \bar{r}_j Q_{jt} \right] (1 + \rho)^{-t}$$

5. Taxable Industrial Property (TIP)

$$TIP_{kt}^{gv} = T_k^g a_t^v \gamma (1 - \theta) I_{kt}$$

where: I_t = investment in period t in county k

θ = allowed deduction from investment (assumed equal to 25 percent)

γ = assessment rate on allowed investment (assumed equal to 40 percent)

a_t^v = coefficient to convert assessed value to taxable value for variant v. Three variants were considered, $a_t = 0.30$ for class 4 property; $a_t = 0.25$ for an equal mix of class 4 and class 2 property; and $a_t = 0.07$ for $t \leq 3$ and $a_t = 0.30$ for $t > 3$ with new industry investments

T_k^g = county levy on taxable industrial property. It is assumed equal to that for mine property. Three variants (g) are considered ($T_k^g = 0.027, 0.031,$ and 0.036)

The proceeds for this tax are derived as a fraction of the taxable value of gross investment with allowance for the class of the investment and the relationship between taxable value and assessed value.

The proceeds of the state levy are derived by summing the taxable value over the counties and applying the state levy as below.

$$TIP_t^v = T^s a_t^v \gamma (1 - \theta) \sum_{k=1}^{S} I_{kt}$$

The county tax proceeds are allocated to a service category by applying the distribution weights used in the case of the county proceeds from the mine property tax and the real and personal property and are given as follows:

$$ATIP_{kst}^{gv} = d_{ks} T_k^g a_t^v \gamma (1 - \theta) I_{kt}$$

where: d_{ks} = distribution weight for county k and service category s

$ATIP_{kst}^{gv}$ = the amount of TIP_k allocated to category s for county k in the tth period with variants designated by the g and v values for the tax levies and taxable values coefficients respectively

6. Electrical Producers' License Tax (EPLT)

$$EPLT_t^q = e \sum_{u=1}^{U} O_t^q \mathrm{kW}_{ut}$$

where: e = tax rate (specified as 1.688 percent of gross income)
    U = number of firms generating electric power
    O_t^q = cost per kilowatt in year t. This is defined to be a function of the fuel cost, interest charges, investment per kilowatt and production expenses. q designates a set of variants defined by assumptions on the interest rate and investments required per kilowatt. Three variations in each factor were considered, so that $q = 1, 2, \ldots, 9$.
    kW_{ut} = kilowatts of generating capacity for firm u in period t

This equation indicates that the state revenue from the electrical producers' license tax is defined by a fraction (e) of the estimated gross income of all producers in the state in each period. The gross income is estimated as the product of the cost per kilowatt (assumed constant over fiirms) and the kilowatts per year for each firm.

The present value of this tax revenue is the sum over time of these payments.

$$PV\text{-}EPLT^q = \sum_{t=1}^{T} \left[eO_t^q \sum_{u=1}^{U} \mathrm{kW}_{ut} \right] (1 + \rho)^{-t}$$

If we consider the present value of a set of projects with variable starting points and project lives, we have:

$$\bar{P}V\text{-}EPLT = \sum_{n=1}^{N} \sum_{t=X_n}^{T_n+X_n} [eO_{nt}\mathrm{kW}_{nt}](1 + \rho)^{-t}$$

where: kW_{nt} = kilowatts provided by project n in period t
    X_n = year in which nth project goes into operation
    T_n = project life in years

7. State Personal Income Taxes (SPIT)

$$SPIT_t = \sum_{h=1}^{H} ATP_{ht} \sum_{f \in A_h} \widehat{EM}_{ft}$$

where: ATP_{ht} = average state income taxes paid by income class h in period t
    H = number of income classes
    $\widehat{EM}_{ft}$ = estimated increment to employment in job category f in period t (an output of the Harris model)

A_h = the set of job categories whose members have average annual income falling within income class h

This equation suggests that we aggregate the increases in employment by job category into classes based on estimates of the average annual income of the members and then apply to the estimated increases in employment now classified by income class the average annual state income taxes paid by each class.

Summing these contributions over time (t) and discounting (with ρ = discount rate) yields the present value of the increments to state personal income taxes as a result of the changes in employment estimated from the Harris model.

$$PV\text{-}SPIT = \sum_{t=1}^{T} \left[\sum_{h=1}^{H} ATP_{ht} \sum_{f \in A_h} \widehat{EM}_{ft} \right] (1 + \rho)^{-t}$$

$PV\text{-}SPIT$ = present value of the state personal income taxes
T = time period for calculation of present value

8. Real and Personal Property Tax (RPPT)

Urban: $RPPT_{kt}^{x} = \bar{T}^{x} \sum_{m \in \Phi_k} R_m \beta_{kt} \widehat{POP}_{kt}$

where: $\bar{T}^{x}$ = tax levy for cities on real and personal taxable property for variant x. Three variants are considered with values corresponding to 0.048, 0.057, and 0.066 of taxable value in dollars

R_m = ratio of real and personal taxable property per thousand population

Φ_k = the set of cities in county k

β_{kt} = fraction of the population living in urban areas in county k and period t

$\widehat{POP}_{kt}$ = estimated population for the county in period t (from the Harris model)

For the case of real and personal property taxes, the tax proceeds are identified according to whether they accrue to urban areas in a given county or to the county government. RPPT for the urban areas are estimated by applying a given tax levy (one for each of three variants) to the tax base. The base is estimated by multiplying the ratio of real and personal taxable property per thousand population for urban residents with the estimated urban population.

Rural: $RPPT_{kt}^{x} = \bar{\bar{T}}^{x} \sum_{n \in \mathfrak{E}_k} R_n (1 - \beta_{kt}) \widehat{POP}_{kt}$

where: $\widetilde{\widetilde{T}}^x$ = tax levy for rural areas of county for real and personal taxable property for variant x

R_n = ratio of real and personal taxable property per thousand population in rural areas

ϵ_k = the set of rural areas in county k

To allocate these tax revenues by functional category, a set of distributional weights are applied to these tax revenues generated in the city and rural areas.

Urban: $ARRPT^x_{kqt} = b_q \bar{T}^x \sum_{m\epsilon\Phi_k} R_m \beta_{kt} \widehat{POP}_{kt}$

Rural: $ARPPT^x_{kst} = d_{ks} \widetilde{T}^x \sum_{n\epsilon\epsilon_k} R_n (1 - \beta_{kt}) \widehat{POP}_{kt}$

where: b_q = fraction of urban tax revenue allocation to category q

d_{ks} = fraction of rural tax revenue allocated to category s in county k. Note this scheme corresponds to that used for taxable industrial property.

Glossary of Terms

e = tax rate for electrical producers' license tax

U = number of firms generating electric power

O_t = cost per kilowatt in year t

kW_{ut} = kilowatts of generating capacity for firm u in period t

kW_{nt} = kilowatts provided by project n in period t

X_n = year in which nth project goes into operation

T_n = project life in years

R = number of firms mining coal

δ = severance tax rate

Ω_{ki} = set of mines owned by firm i located in county k

P_{jt} = price per ton of coal for mine j in period t

Q_{jt} = quantity in tons of coal from mine j in period t

D = allowed deduction in tons

P^*_i = maximum of P_{jt} for all mines owned by a given firm

W_k = county k's required fraction of severance tax deduction

S = number of counties

θ_{lt} = fraction of severance tax receipts allocated to expenditure category l in period t

ψ_t = set of expenditure categories defined in period t to return to the county

I_{kt} = investment in period t in county k

θ = allowed deduction from investment

γ = assessment rate

a_t^v = taxable value/assessable value coefficient

T_k^q = county level on taxable industrial property

d_{ks} = distribution weight for county k and service category s

H = number of income classes

$\widehat{EM}_{ft}$ = estimated increment to employment in job category f in period t

A_H = set of job categories whose members have average annual incomes falling within income class k

r = percentage of gross receipts which function as taxable value for mine property tax

α = indemnity trust account tax rate

K = fixed charge for indemnity trust account tax

η = states' share of royalties

$\bar{r}_j$ = royalty rate for mine j

B = set of mines on federal lands

$\bar{T}^x$ = tax level for cities on real and personal taxable property for scenario x

R_m = ratio of real and personal taxable property per thousand population

Φ_k = set of cities in county k

β_{kt} = fraction of the population living in urban areas in county k and period t

$\widetilde{\bar{T}}^x$ = tax levy for rural areas of county for real and personal taxable property for scenario x

ϵ_k = set of rural areas in county k

b_q = fraction of urban tax revenue from real and personal taxes allocated to category q

5 State Tax Revenue from Montana Coal Development

INTRODUCTION

There are two basic sources for the increased income that the public sector of Montana's economy can expect from coal development. The first, and most significant producer of state revenues, are the types of taxes that apply directly to the mining and sale of coal and converted coal products. The most significant of these is the severance tax, by all odds the most productive source of revenue associated with coal extraction. In addition, there is the state's share of the royalties collected from leases granted private operators to take coal from the federal lands. Perhaps less directly, but closely associated with the development of coal, are the electrical generating facilities based on Rosebud County coal in the vicinity of Colstrip. The electrical producers' license tax will become a significant producer of state tax revenue in the area as the first two large units at Colstrip begin production. Moreover, additional electric power may need to be provided to replace the energy that is now imported from Canadian gas and oil fields, deliveries of which are being phased out.

The second source of state revenue associated with coal development relates to several disparate indirect taxes. First there is the state's share of property tax revenue. The state collects 6 mills on the taxable value of the mines, capital improvements and other taxable industrial property, and on the increment in residential real and personal property associated with the increase in population induced by the coal development.[1] Perhaps a greater source of revenue indirectly related to coal mining is the increment in personal income subject to Montana state income tax, that is, the increase in wages and salaries resulting from expansion in coal extraction and conversion. In this chapter we shall review each of these sources of revenue and make estimates of the yield of the various applicable taxes for each of the scenarios we postulate.

DIRECT TAXES ON VALUE OF COAL AND CONVERSION PRODUCTS

Severance Tax Yields

Of the direct taxes on the value of coal or converted coal products, the severance tax is by far the greatest producer of state tax revenue. This tax, for the range of Btu content of coals with which we are dealing, will be 30 percent of the gross value of the coal produced, value being defined as described and illustrated in chapter 4. In table 5-1 are shown the tonnages that are expected to be produced under scenarios I and II in each county, and the tonnages postulated for scenario III. (It should be recalled that the difference between scenarios III and IV is not in the amount of coal extracted, but rather in the amount converted at site in lieu of export, and thus IV is not shown in the table.)

The Decker and Westmoreland mines are the scenario I and II mines in Big Horn County. The value of their output per ton for tax purposes is $5 and $4 respectively. The Rosebud County mines are the Peabody and Western Energy properties with prices per ton for tax purposes of $2.36 and $3.17 respectively.[2] We elect to use constant price levels as a conservative measure. That is, we seek to avoid introducing purely monetary phenomena such as would be involved if the rise in the general price level were included. On the other hand, some changes in prices are due to changes in the structure of demand. If the price of coal, say, rises *relative*

---

[1] The bulk of the revenue from real and personal property taxes, of course, is reserved for county and city government use and to finance the bulk of the educational expenses associated with school districts.

[2] All prices and costs for this and subsequent chapters will be given, to the extent possible, in dollars of 1974 value. Where this is not possible, or even when possible, where the use of constant 1974 dollar values will bias the results, the implications of the distortion will be evaluated.

Table 5-1. Projected Montana Coal Production, Big Horn and Rosebud Counties, Scenarios I–III

(thousands of tons)

	1975	1976	1977	1978	1979	1980	1981	1982	1983	1984	1985
Scenario I											
Big Horn	12,250	12,250	12,250	12,250	12,250	12,250	12,250	12,250	12,250	12,250	12,250
Rosebud	8,230	8,230	8,230	8,230	8,230	8,230	8,230	8,230	8,230	8,230	8,230
Total	20,480	20,480	20,480	20,480	20,480	20,480	20,480	20,480	20,480	20,480	20,480
Scenario II											
Big Horn	12,250	14,900	17,400	21,000	18,600	20,400	21,000	21,000	21,000	21,000	21,000
Rosebud	8,230	13,430	16,500	17,900	20,700	22,100	22,100	22,100	22,100	22,100	22,100
Total	20,480	28,330	33,900	38,900	39,300	42,500	43,100	43,100	43,100	43,100	43,100
Scenario III											
Big Horn	12,250	14,900	17,490	22,600	23,590	28,400	35,900	37,400	41,000	41,000	41,000
Rosebud	8,230	13,520	18,190	24,490	33,600	43,490	50,100	57,000	58,500	62,100	62,100
Total	20,480	28,420	35,680	47,090	57,190	71,890	86,000	94,400	99,500	103,100	103,100

Source: Scenarios I and II, Montana Energy Advisory Council Coal Development Information Packet, December 1974, Table 12, p. 28. Scenario III, postulated.

to the prices of goods and services in general, this should be reflected in the estimates of value.[3]

As a matter of fact, there are at least a couple of reasons for believing the price of Northern Great Plains coal *may* rise in this fashion. In the first place, it has already risen over the period 1967–74, probably as a result of competition for scarce supplies of low-sulfur coal occasioned by the Clean Air Act of 1967 and subsequent amendments. Now, if this were the only influence on future coal prices, we might expect little or no additional increase or even possibly a decline as increased supplies become available with larger scale and more efficient methods of mining. But coal prices are also likely to be affected by developments in related energy markets. In particular, the dramatic increase in oil prices in 1974 and 1975, described in chapter 1, is beginning to be reflected in coal contracts. Even at the high current price of $5 per ton for western coal, though, the price per Btu is still far below that of energy derived from oil. Of course, we would not expect the Btu price to be exactly the same for coal and oil. As noted in chapter 1, coal is not a perfect substitute for oil in all of its uses, nor indeed in any of its uses in the short run. And there are transport costs to consider. Nevertheless, we would expect some movement in coal prices in response to the oil price increases—though should there be a substantial break in the world oil cartel, falling oil prices would presumably put pressure on the recently established higher coal prices.

We have suggested two reasons why Montana coal prices could rise over the next couple of decades: increased demand for low-sulfur coal, and monopoly in the world oil market. Nevertheless, in making revenue projections based on coal prices, we shall be very conservative. We use the price data provided the Montana Department of Revenue for 1974 by the mine operators themselves for existing and contracted output in scenarios I and II. And for the output from the six new mines in scenarios II and IV we assume a price of $5 per ton, with plus or minus 25 percent for possible sensitivity analysis. We take this conservative approach because of the concerns that have been expressed about the adequacy of coal tax revenues to finance expansion in local public services. If it turns out that revenues appear adequate for this purpose, even assuming a price for

---

[3] Similarly, if the rise in the general price level is reflected in increases in *money* (although not real) personal income, a result will be that individuals having a constant real income will be projected into higher marginal income brackets for tax purposes. We need to be aware of this. We may not be able to make quantitative adjustments to compensate, but we will know the direction of change, and can in any event obtain a lower bound estimate that results from unadjusted estimates.

Table 5-2. Estimated Big Horn and Rosebud County Severance Tax Yields, 1975–90

($000)

Year	Scenario I	Scenario II	Scenario III	Scenario IIIa (−25%)	Scenario IIIb (+25%)
1975	24,186	24,186	24,186	24,186	24,186
1976	24,186	33,106	33,244	33,218	33,270
1977	24,186	39,025	41,692	41,048	42,337
1978	24,186	45,757	58,099	54,968	61,051
1979	24,186	44,820	71,592	64,921	78,263
1980	24,186	48,851	92,873	81,889	103,857
1981	24,186	48,851	113,138	97,089	129,189
1982	24,186	48,851	125,738	106,538	144,938
1983	24,186	48,851	133,388	112,276	154,501
1984–90	24,186	48,851	138,788	116,326	161,251

coal that is unrealistically low, at least for the latter part of our study period, we have convincingly addressed these concerns. If revenues are not adequate, then we can worry about refinements in the assumptions about future coal prices. Given these basic assumptions with respect to the treatment of prices, table 5-2 provides the estimated yield from the state severance tax for each of the coal extraction scenarios.

Although the severance tax is collected by the state, the yield is only partially retained by the state for general state government purposes. Forty percent of the collected revenues are available to the general fund. Ten percent are made available to the school equalization fund. In addition, a potentially variable amount is retained by the state to develop an endowment for future educational purposes. In the early years (1975–79) no less than 10 percent will be made available out of the 27.5 percent earmarked for "local impact and educational trust account" purposes. That is, up to 17.5 percent may, but is not required to be used to support local impact mitigation, with the remainder going into an educational endowment fund. While this could be in excess of the 10 percent required during the first five-year period of the severance taxes operation, this would occur only if the Coal Board allocations for local impact mitigation were to be less than the 17.5 percent maximum allowed under the legislation. Beginning with the year 1980 and thereafter, a minimum of 20 percent (out of a share that is increased to 35 percent of total severance tax receipts after 1979) would be earmarked for the educational trust account, and this could be exceeded, depending on whether the allocation for local impact mitigation were to fall below 15 percent of the severance

Table 5-3. Estimated Big Horn and Rosebud County Severance Tax Revenues for State Government Purposes
($000)

Year	General fund	School equaliza- tion	Educa- tional trust account	Alter- native energy research	Renewable resource develop- ment	Parks acquisi- tion
			Scenario I			
1975	9,674	2,419	2,419	605	605	605
1976	9,674	2,419	2,419	605	605	605
1977	9,674	2,419	2,419	605	605	605
1978	9,674	2,419	2,419	605	605	605
1979	9,674	2,419	2,419	605	605	605
1980–90	9,674	2,419	4,838	967	605	1,209
			Scenario II			
1975	9,674	2,419	2,419	605	605	605
1976	13,242	3,311	3,311	828	828	828
1977	15,610	3,903	3,903	976	976	976
1978	18,303	4,576	4,576	1,144	1,144	1,144
1979	17,928	4,482	4,482	1,121	1,121	1,121
1980–90	19,540	4,885	9,770	1,954	1,221	2,443
			Scenarios III and IV			
1975	9,674	2,419	2,419	605	605	605
1976	13,298	3,324	3,324	831	831	831
1977	16,677	4,169	4,169	1,042	1,042	1,042
1978	23,204	5,801	5,801	1,450	1,450	1,450
1979	28,637	7,159	7,159	1,790	1,790	1,790
1980	37,149	9,287	18,575	3,715	2,322	4.644
1981	45,255	11,314	22,626	4,526	2,828	5,657
1982	50,295	12,574	25,148	5,030	3,143	6,287
1983	53,355	13,339	26,678	5,336	3,335	6,669
1984–90	55,515	13,879	27,758	5,552	3,470	6,939

tax yield. İn column 4 of table 5-3 we assume as a practical matter that 10 percent of the severance tax yields is destined to go to the educational trust account during 1975–79, and 20 percent thereafter, recognizing that these are minima which could be exceeded.

During the first five years of its operation (1975–79), 7.5 percent and thereafter 11.5 percent are earmarked for a combination of (a) alternative energy research (2.5 percent and 4 percent), (b) renewable resource development (2.5 percent and 2.5 percent), and (c) park acquisition (2.5 percent and 5 percent), as described in chapter 4. The sums that are available for state governmental purposes corresponding to each scenario then are shown in table 5-3.

The difference between the sum of funds available to the several ex-

Table 5-4. Estimated Tonnages of Federal Coal from Big Horn and Rosebud Counties, 1975–90

(thousands of tons)

Year	Scenario I	Scenario II	Scenarios III & IV
1975	4,916	4,916	4,916
1976	4,916	7,857	7,947
1977	4,916	9,425	11,205
1978	4,916	11,877	18,627
1979	4,916	12,455	30,345
1980	4,916	14,427	43,817
1981	4,916	14,427	57,327
1982	4,916	14,427	69,327
1983	4,916	14,427	70,827
1984–90	4,916	14,427	74,427

penditure accounts in table 5-3 and total yield of the tax given for each scenario in table 5-2 represent the funds that are earmarked for special county and local area purposes such as highway improvement, local educational requirements, and related special purposes arising out of the impact of coal development at the local level. The funds available to county, town, and school districts will be discussed in chapter 6 where we take up the matter of revenues associated with coal development that accrue to the lower level taxing jurisdictions.

State's Share of Royalties

The pattern of private, state, Indian, and federal ownership of the lands overlying the coal beds is one of substantial intermingling. A mining operation that has federal leases will very likely be mining from private or nonfederal public lands as well. The proportion of the coal that may come from federal lands is likely to vary from month to month in the case of some of the operators, but taking a longer view, and an estimated "average," we can get at least some approximation of the amount of coal that is mined which will provide a share of the federal royalties for the state.[4] From information supplied by the U.S. Geological Survey in the field, the data in table 5-4 represent an estimate of the amount of federal coal mined under each of the scenarios. Roughly 35–40 percent of Peabody Coal Company's 3 million tons per year is estimated to fall in this class in the future. Approximately a third of the projected increase in coal

[4] Obtained from personal communication with Douglas Hileman, U.S. Geological Survey, Billings, Montana, who is the local representative of USGS entrusted with monitoring federal leases in this regard.

Table 5-5. Estimated States' Share of Federal Royalties from Big Horn and Rosebud Counties, 1975–90

($000)

Year	Scenario I	Scenario II	Scenarios III & IV	Scenario IIIa (−50%)	Scenario IIIb (+50%)
1975	323	323	323	323	323
1976	323	516	544	527	569
1977	323	619	1,175	828	1,662
1978	323	779	2,888	1,570	4,734
1979	323	817	6,408	2,913	11,299
1980	323	947	10,131	4,391	18,168
1981	323	947	14,353	5,974	26,084
1982	323	947	18,103	7,381	33,115
1983	323	947	18,572	7,556	33,994
1984–90	323	947	19,697	7,978	36,103

production from Western Energy mines, from a 1975 level of 5.2 million to 19 million by 1980, is expected to come from federal coal on which royalty payments are required. Currently, about a quarter of the Decker mine production of 8.25 million tons in Big Horn is federal coal. Expansion by 1980 to approximately 14 million tons will see approximately a half of the total represented by federal coal. The Westmoreland mine, on the other hand, will take coal to which the Crow Indians hold rights, with no federal coal involved. Moreover, a large coal operation under consideration by Shell also will involve virtually no federal coal. However, most of the remainder of the increment in coal production is expected to involve federal leases on which royalty payments are due. Accordingly, in table 5-5 we present the share of the royalty the state is due during the period 1975–90, to a first approximation based on the best estimates of persons working closely with the problem.

The state's share of the royalties paid on coal mined from federal lands is based on the best estimates of the mining plans of current coal operators. In scenario III, however, we have postulated additional development which has not yet been specifically identified. Accordingly, the estimates for scenario III are supplemented with alternatives IIIa and IIIb which assume that the scenario III estimate of federal coal tonnages might respectively be (a) reduced by 50 percent, and (b) increased by 50 percent, to test our results for sensitivity to the assumptions made about the mix of federal coal in the future total of coal production.

A word is in order concerning the *ad valorem* rates used for estimating royalties. As discussed in chapter 4, currently operating leases require

Table 5-6. Estimated Yields from the Resource Indemnity Trust Account Tax, 1975–90

($000)

Year	Scenario I	Scenario II	Scenario III	Scenario IIIa (−25%)	Scenario IIIb (+25%)
1975	409	409	409	409	409
1976	409	557	560	559	560
1977	409	656	701	689	712
1978	409	768	973	922	1,024
1979	409	753	1,200	1,088	1,312
1980	409	820	1,555	1,371	1,738
1981	409	820	1,892	1,624	2,160
1982	409	820	2,102	1,782	2,423
1983	409	820	2,230	1,877	2,582
1984–90	409	820	2,320	1,945	2,695

royalty payments as stipulated in the Mineral Leasing Act of 1920. These apply largely to coal that is extracted from the federal lands appearing in our scenarios I and II. For scenario III with the 1976 amendment to the Act, the increased *ad valorem* rate and increased state's share are reflected in a severalfold increase in the royalties compared with yields under the previous leasing terms. It should be mentioned as well that a part of the gain in projected yields to the state from this source is accounted for by the higher average price per ton in scenario III than is reflected in the coal contracted for in scenarios I and II.

Resource Indemnity Trust Account

The last tax we discuss that is levied directly on the coal mined is the resource indemnity trust account tax. The law exempts the first $5,000 in gross value of output and taxes everything in excess of that amount at 0.5 percent. In addition, a fixed amount of $25 per mine is levied. The intent of the legislation appears to be that any mine operator with a gross income of less than $5,000 is exempt, with every operator having a gross income in excess of $5,000 being taxed at a rate of 0.5 percent on the *total* value of output. That is, the $25 constant added to the sum figured at 0.5 percent of everything in excess of $5,000 figures to be exactly 0.5 percent on the first $5,000 as well. For our purposes, then, the resource indemnity trust account tax will be simply 0.5 percent on the value of total output from each mine in each county since we will be working with large volumes and values of mine output. The sums yielded by this tax are shown in table 5-6.

Table 5-7. Estimated Yield to State of 6-Mill Levy on Taxable Mine Property as Determined by the Gross Proceeds Method
($000)

Year	Scenario I	Scenario II	Scenario III	Scenario IIIa (−25%)	Scenario IIIb (+25%)
1975	218	218	218	218	218
1976	218	299	300	300	300
1977	218	352	376	370	382
1978	218	413	523	496	551
1979	218	404	646	585	706
1980	218	440	837	738	936
1981	218	440	1,020	875	1,164
1982	218	440	1,133	960	1,306
1983	218	440	2,202	1,011	1,392
1984–90	218	440	1,250	1,048	1,453

Gross Proceeds Tax

In discussing taxes that are levied directly on the volume or value of coal produced, we might think also of the gross proceeds tax, although this is rather a method of determining the taxable value of the mine—and by mine we mean the excavation itself, rather than structures and improvements above ground associated with the mine workings. But the method involves a procedure that depends on the value of the mine output to establish the taxable base, against which are applied the millage levies of the several taxing jurisdictions to provide real property tax revenues.

The state of Montana levies 6 mills against real and personal property. Accordingly, the taxable value of the mines determined by the gross proceeds method will produce revenues for the state as given in table 5-7.

The gross proceeds are reckoned in terms of 1974 prices as reported to the Montana Department of Revenue for tax purposes, and in accord with our practice of presenting data in terms of 1974 dollar values, we retain these for all mines that are in operation in scenarios I and II. In scenario III, where we include mines postulated under an accelerated rate of development, a value of $5 per ton has been assumed. To indicate the sensitivity of the results in scenario III to the assumption as to price, in IIIa and IIIb we display the revenues alternatively on the assumption that the prices are 25 percent less and 25 percent higher than the base $5 price, respectively.

Taxable Industrial Property

The gross proceeds tax, as indicated, simply provides a procedure used to determine, in a manner of speaking, the "taxable value of the coal" in a working mine. The above-ground improvements, or the investment in complementary facilities, represents taxable industrial property. The method used, as described in detail in appendix A, chapter 4, involves taking 40 percent of the "market value" of the investment (operationally determined to be three-quarters of the investment) as the assessed value, with 30 percent of the assessed value serving as the taxable value. In practice, the problem may become somewhat complicated by virtue of the fact that much of the investment in a truck and shovel strip-mining operation may be the investment in trucks that could be interpreted to fall within class 2 property, the taxable value of which is taken to be only 20 percent of assessed value. If we take the investment to be divided equally between the mobile equipment qualifying for the 20 percent rule, the taxable value would average then about 25 percent of the assessed value for a mine. Similarly, if we consider the power and gasification plants as industrial property, recognizing that these will have a significant investment in pollution abatement facilities which are allowed to enter the tax rolls at only 7 percent of assessed value, an adjustment in the computation will be required on this score as well. Polzin cites Montana Power and Light Company's estimate of investment in pollution abatement facilities running to 20 percent of its total investment.[5] If we take this as a representative proportion of investment qualifying under the 7 percent rule, with the remainder taken at 30 percent, the weighted average applicable to assessed value in order to determine taxable value comes out, rounded, to 25 percent. Accordingly, we will assume that the taxable value will be given as $(0.25)(0.4)(0.75)$ (investment), or in other words as 7.5 percent of the value of the investment, with the state's receipts being computed at 6 mills of the taxable base.

For purposes of sensitivity tests, however, the estimates have been made giving the estimated tax yield based both on the disallowance of the class 2 qualification for the type of very large hauling equipment associated with strip mining, and similarly on a straight 30 percent applicable to the total of industrial property. As a low yield estimate we assume, alternatively, that the mining operations and energy conversion will qualify as "new enterprises" for the state and will, therefore, qualify for the 7 percent credit applicable to new enterprises during the first three years of operation. The medium estimated state tax yields are shown in table 5-8, with the high and low estimates reflected in the total upper and lower bound estimates of the summary tables.

---

[5] Paul Polzin, *Water Use and Coal Development in Montana,* Montana University Joint Water Resources Research Center, Bozeman, Montana, 1974.

Table 5-8. Estimated Annual Yield from Industrial Property Tax
($000)

Year	Scenario I	Scenario II	Scenario III	Scenario IV	Scenario IVa
1975	35	35	35	35	35
1976	134	147	148	148	148
1977	134	157	167	167	167
1978	134	165	203	203	203
1979	134	170	242	242	242
1980	134	172	271	271	271
1981	134	172	297	297	297
1982	134	172	317	805	805
1983	134	172	323	812	812
1984	134	172	326	815	815
1985–87	134	172	329	1,360	871
1988–90	134	172	329	1,537	1,537

Electrical Producers' License Tax

Montana requires payment of 1.688 percent of the gross value of electricity generated as a tax on electrical production. The gross value, of course, is given by the rates which an electric utility is permitted to charge by the public utility commission. These in turn are based on an attempt to provide a "fair return" on investment. For this purpose, an investment "rate base" is determined on which a "fair," or competitive, rate of return is permitted and a schedule of electric rates authorized to provide this rate of return to the utility operating in a protected franchised territory. There are variations among state utility commissions as to what may be included in the rate base. The commission in Montana appears to be among the most generous.[6] It is not possible to determine precisely what the gross value of production from the 700 MW of capacity represented by Colstrip units 1 and 2 will be. Yet, taking everything into account, one can come to an estimate good to a first approximation that is in the neighborhood of $71 per kilowatt installed per year.[7] This would yield

[6] Irrespective of what the state public utility commissions judge to represent legitimate investment to provide the "base" on which a rate of return is calculated, the U.S. Internal Revenue Service is guided by its own criteria. In this instance it disallows a substantial amount of what is permitted for inclusion by the state commission, and as a consequence, the Montana Power Company is required to pay one of the highest rates of federal corporate income tax among electric utilities (the rate in this case is figured as the amount of the tax in relation to investment). All of which suggests that the schedule of rates used by the utility, in this case, represents a high gross value relative to utilities in general.

[7] If we take the investment for rate purposes at $350 per kilowatt installed, a reasonable figure at 1974 prices and costs, and assume an interest rate of 9 percent, the fixed charges would run about 15.66 percent or about $54.80 per kilo-

Table 5-9. Estimated Annual Yield from Electrical Producers' License Tax
($000)

Year	Scenarios I, II, III	Scenario IV	Scenario IVa
1975–81	839	839	839
1982–84	839	3,955	3,955
1985–87	839	7,071	3,955
1988–90	839	7,071	7,071

a gross value of electricity production, associated with the 700-MW capacity at Colstrip of $49,700,000 per year. At 1.688 percent, the tax
would produce revenue of about $839,000 for the state each year.[8] Ac-
cordingly, while the tax on electrical generation is not a direct tax on
coal, it is a direct tax on a coal conversion product that has the capacity
for yielding substantial revenue to the state. If, for example, the phasing
out of the export of 50 billion cubic feet of Canadian natural gas continues as anticipated, and the energy is replaced with electricity for domestic (Montana) consumption, a very large increase in the production
of electricity will be required even in the absence of a change in the policy
that restricts coal conversion to the amount required by Montana.[9] In
table 5-9, we present the estimated annual yield from the electrical producers' license tax. The Colstrip power plant (units 1 and 2) is reflected
in scenarios I, II, and III. Scenario IV considers two 2,600-MW power

---

watt per year. Assuming a 9,000 Btu heat rate per kilowatt hour and 8,200 Btu
per pound coal, roughly 1.10 pounds of coal would be required per kilowatt hour.
At a plant factor of 80 percent, some 7,700 pounds, or 3.85 tons of coal would
be burned annually per kilowatt installed, and at $3.17 per ton (the rate Western
Energy reports charging for tax purposes in 1974), the fuel costs would be an
additional $12.20 per kilowatt year. Given another $4 for nonfuel production expenses, we have $71/kW/yr.

[8] A plant factor of 80 percent may be a bit low for the early years, and perhaps high for later years. Over the first 15 years, during which we are computing
revenue yields, however, it may be as good a figure to use as any. If the plant
factor declines, however, we would expect the fuel component, rather than the
capacity component, to decline, and since that represents only the smaller fraction
of the computed total cost (or value) of production, the use of $71/kW/yr is
sufficiently close to give estimates to a first approximation.

[9] 50 × 10^9 cubic feet of natural gas represents at 1,000 Btus per cubic foot,
50 × 10^{12} Btus. With a conversion efficiency of about 33 percent from coal to
electricity, but a somewhat higher efficiency in conversion of Btus in electricity to
work and space heating than in the case of natural gas, one would need about
1.25 × 10^{14} Btus of coal production to compensate. For coal with a Btu content
of 7,500 to 9,000 per pound, we foresee the additional coal extraction of 10 million to 11 million tons per year, and the generation of 2,600 MW of electricity or
thereabouts, to meet this contingency. A tax on such generation would yield upwards of $3 million in state revenues annually.

plants not necessarily consistent with the present policy of restricting energy conversion to meet only the needs of Montana residents and economic activities, the first coming on line in 1982 and the second coming on line in 1985.

Finally, in scenario IVa we consider a slower schedule for the second 2,600-MW power plant, with the latter coming on line in 1988.

INDIRECT TAXES RELATED TO COAL DEVELOPMENT

State Personal Income Taxes

Although indirect taxes such as personal income and the state's 6-mill levy on real and personal property do not yield individually as much in revenue as some of the direct taxes on coal (e.g., severance), or products of coal conversion (electrical producer's license), the personal income tax on the increment in income from employment and earnings in coal mining and induced activities is nonetheless potentially a significant revenue producer for the state. The amount of increased state personal income tax yield, of course, is dependent on the postulated level of development.

It is not possible to compute exactly the increment in personal income tax revenue due to coal development in Big Horn and Rosebud counties, but a reasonably close estimate can be obtained by taking the estimated change in number of jobs in each of the sectors for which we obtain employment data, and given the average earnings per employee in each sector, estimate what the increment in employment and earnings would yield in state personal income tax revenues for each scenario. Since earnings by employees correspond closely to the adjusted gross income, this method will pick up virtually all of the personal income subject to tax that will be associated with the increment in employment due to coal development. It leaves out of account, however, increases in property income, but this omitted item will represent a relatively small, if not negligible fraction of the total increment in personal income due to coal development.

In table 5-10 we give the estimated average tax paid per thousand-dollar income interval in 1974 by Montana residents, as estimated by the Montana Department of Revenue.[10] Combining the estimated number of

[10] Provided by courtesy of John Clark, Administrator, Research Division, Montana Department of Revenue.

Table 5-10. Estimated Individual Income Tax Payments by Income Interval, 1974, Montana Department of Revenue

Adjusted gross income	Average tax liability	Adjusted gross income	Average tax liability
$ 0	$ 0	$25,000–25,999	$1,121
0– 999	1	26,000–26,999	1,237
1,000– 1,999	12	27,000–27,999	1,301
2,000– 2,999	29	28,000–28,999	1,367
3,000– 3,999	51	29,000–29,999	1,460
4,000– 4,999	80	30,000–30,999	1,539
5,000– 5,999	107	31,000–31,999	1,646
6,000– 6,999	138	32,000–32,999	1,692
7,000– 7,999	169	33,000–33,999	1,768
8,000– 8,999	202	34,000–34,999	1,828
9,000– 9,999	232	35,000–35,999	1,900
10,000–10,999	263	36,000–36,999	2,039
11,000–11,999	299	37,000–37,999	1,912
12,000–12,999	337	38,000–38,999	2,130
13,000–13,999	377	39,000–39,999	2,244
14,000–14,999	420	40,000–40,999	2,193
15,000–15,999	468	41,000–41,999	2,403
16,000–16,999	524	42,000–42,999	2,629
17,000–17,999	581	43,000–43,999	2,584
18,000–18,999	638	44,000–44,999	2,565
19,000–19,999	705	45,000–45,999	2,807
20,000–20,999	770	46,000–46,999	2,834
21,000–21,999	839	47,000–47,999	2,734
22,000–22,999	900	48,000–48,999	2,776
23,000–23,999	989	49,000–49,999	2,992
24,000–24,999	1,050	50,000–	4,752

individuals within each thousand-dollar income interval with the estimated average individual income tax paid for each interval gives us the estimates of state personal income tax revenues, by scenario, as shown in table 5-11.

The estimates given in table 5-11 are suspected of being somewhat higher than the yields that are likely to occur in actuality, for reasons alluded to in chapters 2 and 3, and discussed more fully in the following two chapters. We have reason to feel that at the present time (scenario I and II level activities 1975–76), the growth in the population predicted for Big Horn County by our model is somewhat higher than what actually is occurring under scenario I and early scenario II conditions. The reason seems to be that Sheridan, Wyoming, has somewhat better services, particularly educational ones, than are currently available in Big Horn

Table 5-11. Estimated State Personal Income Tax
($000)

Year	Scenario I	Scenario II	Scenario III	Scenario IV	Scenario IVa
1975	2,098	2,098	2,118	2,118	2,118
1976	2,070	2,220	2,339	2,372	2,372
1977	1,947	2,217	2,446	2,680	2,680
1978	1,990	2,359	2,747	3,376	3,376
1979	2,007	2,403	2,944	4,098	4,064
1980	2,054	2,519	3,220	4,653	4,386
1981	2,120	2,640	3,527	5,066	4,334
1982	2,160	2,692	3,621	5,431	4,310
1983	2,183	2,726	3,702	5,888	5,019
1984	2,241	2,787	3,767	5,182	5,268
1985	2,273	2,821	3,795	5,110	5,960
1986	2,374	2,916	3,905	5,099	6,014
1987	2,464	3,065	4,135	5,284	5,705
1988	2,540	3,141	4,215	5,398	5,608
1989	2,570	3,172	4,277	5,436	5,538
1990	2,640	3,249	4,418	5,540	5,629

County, with the result that a significant portion of workers in the Birney–Decker area reside in Sheridan and commute to their place of employment. Whether this situation would persist with the level of expansion corresponding to scenarios III and IV cannot be predicted with certainty. In any event, the personal income tax yields associated with the Big Horn County activities in the Birney–Decker area would overstate the amount going to the state of Montana, at least initially, and we should take note of the fact. On the other hand, state income tax revenues are not available for local impact mitigation in any event except indirectly through a portion available for state equalization of support for education. As we shall see in chapter 7, this will be of little consequence for funding educational services in the coal areas.

Real and Personal Property Tax

While in a sense the revenue raised in connection with the so-called gross proceeds tax, and the tax on industrial property, represents taxes on real estate, the bulk of the real and personal property taxes available to many of the taxing jurisdictions represents nonindustrial, residential, and relatively small commercial properties, and the personal property as-

Table 5-12. Estimated State Real and Personal Property Taxes Exclusive of Mine and Industrial Properties
($000)

Year	Scenario I	Scenario II	Scenario III	Scenario IV	Scenario IVa
1975	93	93	96	96	96
1976	74	87	104	107	107
1977	59	83	108	126	126
1978	56	91	123	173	173
1979	54	90	139	222	220
1980	56	95	157	263	247
1981	53	97	168	286	234
1982	56	101	174	312	235
1983	55	99	176	342	277
1984	59	103	180	296	292
1985	58	103	184	285	331
1986	62	107	183	278	336
1987	64	108	189	247	305
1988	66	113	190	276	294
1989	70	114	197	274	283
1990	72	120	198	280	285

sociated with residential and small-scale commercial real estate. The determination of the taxable base against which property taxes are levied is an extremely involved and arcane business, known only to county assessors, and at times, it is suspected, not even to them. Since the state collects only a fixed 6 mills on the taxable value of real and personal property, we present in this chapter only our estimates of the tax yield to the state, deferring to later chapters a more extended discussion of the nature of the real and personal property tax yields and the determination of the bases on which the taxes are levied. In table 5-12 then, are given our estimates of the state revenues derived from its 6-mill levies against the real and personal property tax base estimated to be associated with each of the coal extraction scenarios.

SUMMARY OF STATE REVENUES FROM MONTANA COAL DEVELOPMENT

Having reviewed each of the several taxing instruments and sources of revenue accruing for state government use, it may be useful to provide a

Table 5-13. Estimated Total Tax Revenues from Coal Development for State Government Purposes

($000)

Year	Scenario I	Scenario II	Scenario III	Scenario IV	Scenario IVa
1975	20,341	20,341	20,364	20,364	20,364
1976	20,392	27,012	27,273	27,309	27,309
1977	20,255	31,264	33,954	34,206	34,206
1978	20,295	36,300	47,453	48,131	48,131
1979	20,309	35,731	60,741	61,979	61,942
1980	23,744	45,645	92,703	94,241	93,958
1981	23,808	45,769	114,303	115,961	115,176
1982	23,851	45,825	128,765	134,318	133,120
1983	23,873	45,857	135,756	141,712	140,779
1984	23,934	45,921	141,493	146,628	146,711
1985	23,966	45,955	141,528	150,206	147,498
1986	24,070	46,055	141,637	150,188	147,557
1987	24,163	46,204	141,872	150,343	147,217
1988	24,241	46,285	141,954	150,662	150,890
1989	24,274	46,317	142,022	150,698	150,809
1990	24,347	46,400	142,164	150,808	150,902

Note: Because of rounding, totals shown above may differ slightly from those obtained by summing the various tax yields.

summary of the tax revenues by scenario and year. Table 5-13 provides such an estimate of revenues accruing to Montana from taxes levied either directly on coal extraction and conversion, or received indirectly from taxable incomes and property values associated with coal-related activities.

Since the estimated tax receipts have been based on various assumptions as to the price of coal, the proportion of the mined coal which comes from federal lands that return a share of royalties, and similar variations that would affect the estimated totals, it is desirable to provide a range of estimates that should bracket the expected estimates for each scenario. To this end we have summed the receipts that would be forthcoming using the smallest estimates given in tables in the text, giving us a lower bound figure for sensitivity analysis. These estimates are shown in table 5-14. A correspondingly constructed higher bound set of estimates for each scenario is given in table 5-15. This set of tables will permit checking the sensitivity of our estimates to the assumptions made in their construction.

Table 5-14. Estimated Total Tax Revenues from Coal Development for State Government Purposes (Lower Bound)
($000)

Year	Scenario I	Scenario II	Scenario III	Scenario IV	Scenario IVa
1975	20,297	20,297	20,320	20,320	20,320
1976	20,353	26,959	27,181	27,216	27,216
1977	20,218	31,206	33,084	33,332	33,332
1978	20,284	36,281	43,951	44,620	44,620
1979	20,299	35,713	52,509	53,730	53,694
1980	23,733	45,626	77,675	79,192	78,912
1981	23,797	45,749	92,378	94,011	92,237
1982	23,840	45,804	101,854	107,379	106,197
1983	23,862	45,837	106,956	112,879	111,958
1984	23,923	45,901	110,854	115,966	116,049
1985	23,954	45,934	110,888	119,546	116,829
1986	24,058	46,034	110,997	119,530	116,887
1987	24,150	46,183	111,231	119,712	116,553
1988	24,228	46,263	111,313	120,004	120,228
1989	24,260	46,294	111,379	120,040	120,149
1990	24,332	46,377	111,521	120,149	120,242

Note: Because of rounding, totals shown above may differ slightly from those obtained by summing the various tax yields.

Table 5-15. Estimated Total Tax Revenues from Coal Development for State Government Purposes (Upper Bound)
($000)

Year	Scenario I	Scenario II	Scenario III	Scenario IV	Scenario IVa
1975	20,366	20,366	20,388	20,388	20,388
1976	20,414	27,039	27,347	27,383	27,383
1977	20,274	31,293	34,929	35,184	35,184
1978	20,313	36,331	51,475	52,164	52,164
1979	20,327	35,763	70,355	71,610	71,572
1980	23,763	45,678	110,039	111,599	111,313
1981	23,826	45,803	139,601	141,282	140,487
1982	23,869	45,859	159,998	165,578	164,365
1983	23,891	45,891	169,006	174,995	174,049
1984	23,953	45,957	176,864	182,022	182,104
1985	23,985	45,990	176,900	185,598	182,940
1986	24,090	46,091	177,009	185,579	182,960
1987	24,183	46,240	177,245	185,760	182,613
1988	24,261	46,322	177,327	186,053	186,284
1989	24,295	46,354	177,397	186,088	186,201
1990	24,368	46,439	177,539	186,200	186,295

Note: Because of rounding, totals shown above may differ slightly from those obtained by summing the various tax yields.

6 Coal-Related Tax Revenues of Local Governments

INTRODUCTION

For all practical purposes, both the source of tax revenues and the demand for the public services associated with coal extraction and related developments occur mostly within the state.[1] But as we go increasingly lower in the hierarchy of local governments, we can expect to encounter increasing dissociation between the location of the taxable properties corresponding to the activities in question and the place, and hence taxing jurisdiction, where demands for public services may be felt. For example, it may be expected that the site of a coal mine or conversion facility would normally be located somewhere outside of town. The population brought in by the new industry, however, may live predominantly in one or another of the towns or cities where service facilities already exist or where they may be expanded with relatively greater ease than in the rural tax jurisdiction in which the coal and related properties are located.

It is partly for this reason that the state severance tax revenues are earmarked for specific purposes destined, in part, for the jurisdictions where the demand for public services will be *felt,* whether or not the tax-

---

[1] There is evidence that much of the work force that is involved with the expansion of output from the coal fields in the Birney–Decker area of Big Horn County, however, is being domiciled, initially at least, in the Sheridan, Wyoming, area.

able property is located within the jurisdiction. Similarly, 25 percent of the state's personal income tax is earmarked for school equalization aid in order to assure a given basic level of education for pupils in every school district, irrespective of the value of taxable property in the district. Although these funds may be available in part to the school districts in the coal areas, it is not likely that a contribution proportionate to personal income tax payments made from these areas will be returned to them.[2]

In spite of mechanisms designed to help equalize the capacity for supplying local public services, communities still often depend on the property tax as the major source of revenue to finance these services. The distribution of taxable property associated with large industrial undertakings in rural areas is very unequal across local taxing jurisdictions. This frequently poses serious problems for local units of government Given the possibility, if not the likelihood, of an unequal fiscal burden associated with any rapid and extensive expansion of coal extraction in Montana, it is desirable to evaluate the tax revenues that would be available to various units of local government for later comparison with the expenditures that these units might be called on to make.

Accordingly, in this chapter we attempt a first approximation overview of the tax receipts that would be available to counties and other levels of local government. It should be noted that beyond the level of the county government, the revenues to units of local government become less firmly established. This is in part because on the one hand revenues of specific school districts and even towns are partly dependent on the discretion of special fund disbursement boards, such as the coal board, set up to oversee the distribution of earmarked severance and other tax receipts. These sums are not automatically fixed for each taxing jurisdiction in which demands for services will be felt. For example, a significant part of the population associated with the development in the Colstrip area of Rosebud County is domiciled in Forsyth, where much of the increase in the demand for community services associated with the Colstrip development will be felt. A mandatory minimum levy of 40 mills is required of every school district to provide for the "foundation" program, a state-mandated minimum standard. A levy of this amount on the taxable property,

---

[2] The wage and salary incomes of workers in coal and coal-related activities are substantially higher than the average across all economic activities, and these in turn will be available for distribution across the state, rather than just the counties in which the coal activity takes place. But even more significant is the relation between the local tax base available to support the educational foundation program and the funds required for this purpose. This matter will be discussed more fully in the following chapter.

mines, and power plants associated with Colstrip (school district 19) will yield a substantially greater amount of revenue than is required to meet the elementary and high school foundation programs in that district. The excess will go to the county school equalization fund for distribution to all of the districts within the county to enable them to meet their foundation level programs.[3] But an investigation would be required to determine whether these receipts will suffice to meet the tax base deficiency of each affected area. Moreover, the foundation program is generally substantially below the standard of education that is provided by the better endowed towns and cities. It requires some study, then, to determine whether proceeds of the severance tax potentially available for return to the affected areas in the coal counties are equal to the demand arising from expectations of educational standards that exceed the minimum foundation program. We address this matter directly in chapter 7.

In the next section we identify the several sources of revenues available to local governments and provide estimates of the yield of each by county for the coal development scenarios.

DIRECT TAXES ON VALUE OF COAL AND CONVERSION PRODUCTS

Severance Tax Yields

The severance tax on coal is a state tax as described in chapter 4, but the yields from this tax are earmarked partly for various state and local governmental purposes. A substantial portion [17.5 percent of the severance tax yield for the first five years of its operation as a maximum (1975–79) and 15 percent thereafter][4] is available to defray costs of mitigating local impacts, and is available to provide somewhat better facilities and programs than can be provided for by the foundation program alone. Such monies, however, cannot be used in lieu of levies in support of the foundation program. Additional categories earmarked for local governmental purposes include 10 percent of the severance tax yield allocated to improving highways in the area where the coal is mined, for a period of five years (1975–79) following enactment of the severance tax; a 4 percent allocation for general fund purposes to the county where the coal is

---

[3] These funds are available to defray operating expenses only and are not available for expansion of facilities.

[4] Recall the discussion in chapter 5 of the potentially variable nature of the severance tax percentage allocation to mitigate local impacts.

Table 6-1a. Big Horn County Severance Tax Yields for Local Government Purposes, 1975–90

($000)

Year	Scenario I	Scenario II	Scenario III	Scenario IIIa (−25%)	Scenario IIIb (+25%)
1975	5,564	5,564	5,564	5,564	5,564
1976	5,564	6,856	6,856	6,856	6,856
1977	5,564	7,831	7,883	7,874	7,891
1978	5,564	9,586	10,374	10,182	10,567
1979	5,564	8,416	10,847	10,244	11,450
1980	3,167	5,290	7,509	6,957	8,061
1981	3,167	5,290	9,424	8,393	10,455
1982	3,167	5,290	9,840	8,705	10,975
1983–90	3,167	5,290	10,839	9,454	12,224

mined (3.5 percent after the first five years); and 1 percent of the severance tax yield for county planning purposes for the initial five-year period.

If we assume that the percentages of the severance tax collected from the coal workings of each county that are earmarked for local government purposes are returned to the county of origin,[5] and we take the estimated tonnages produced by each county as shown in table 5-1 at the prices described there, we obtain estimates of the funds raised by the severance tax that are available to each county for each scenario. These are shown in tables 6-1a and 6-1b.

Gross Proceeds Tax

The revenues available for local government use from the so-called gross proceeds tax depend on the levies that will be applied by the taxing jurisdictions within which the mines are located, as well as on the formula for determining the taxable property value. The gross proceeds method of determining the property value subject to taxation involves taking 45 percent of the value of the output of the mines as the taxable value of the mine exclusive of the investment in above-ground facilities. To obtain the estimated value of coal production, as in the case of the severance and gross proceeds tax discussions in chapter 5, we use the actual prices reported by the mine operators to the Montana State Department of Revenue for mines in operation during 1974, and we assume that the price per ton of coal (in 1974 dollars) that will come from mines that are brought into production after 1974 in each of the scenarios will be $5

---

[5] This assumption is made for computational convenience, there being no requirement that the funds return to the county of origin.

Table 6-1b. Rosebud County Severance Tax Yields for Local Government Purposes, 1975–90
($000)

Year	Scenario I	Scenario II	Scenario III	Scenario IIIa (−25%)	Scenario IIIb (+25%)
1975	2,296	2,296	2,296	2,296	2,296
1976	2,296	3,903	3,948	3,940	3,957
1977	2,296	4,852	5,667	5,466	5,868
1978	2,296	5,285	8,479	7,683	9,275
1979	2,296	6,150	12,421	10,856	13,985
1980	1,307	3,747	9,672	8,193	11,152
1981	1,307	3,747	11,507	9,568	13,445
1982	1,307	3,747	13,421	11,004	15,838
1983	1,307	3,747	13,838	11,317	16,359
1984–90	1,307	3,747	14,837	12,066	17,608

plus or minus 25 percent. Finally, we examine yields of various postulated prices in order to permit testing how sensitive our conclusions in subsequent analyses might be to the specific estimate of coal prices.

The taxable value determined by the gross proceeds method is relatively simple once the price and tonnage estimates are made. The tax yields, however, depend additionally on what is assumed with respect to the rate at which the property value is to be taxed. In theory, at least, local governments determine the range of public services to be provided through tax revenues and how much of each service is to be provided, along with estimates of their costs. Given the tax base, it only remains to fix the rate to be levied against the taxable value of the property so that the revenues will equal the budgeted expenditures. Counties with equivalent needs in terms of types and amount of public services will need to levy different rates, depending on their aggregate taxable property.

When we consider the tax on mines, it appears that only the county and some rural school districts are likely to have mines and related industrial facilities in their tax base. That is, such mining and industrial activities typically do not take place in cities or towns. For this reason, the tax revenues that would be anticipated would correspond to those that would reflect the millage rates of only counties and school districts.

In the case of the school district having a large mining or coal-related industrial property, revenues that would be yielded with typical tax rates are likely to be very large in relation to public service expenditure requirements in that specific taxing jurisdiction. The temptation is to lower the rate in order to reduce the real and personal property taxes of individual property owners. It is partly for this reason that there is a mandated lower limit to the rate at which a school district's property can be taxed, namely, 40 mills.

But even at 40 mills, where a substantial investment has been made in mining or industrial property, the yield will exceed by a considerable margin the local school district's budget requirements based on a given minimum per student expenditure to provide for the basic educational program. Since all monies raised by the 40-mill levy that are in excess of the requirement to finance the foundation program for the district are available through the school equalization fund for redistribution among county school districts unable to finance the basic foundation program, an affluent school district looking toward an enriched program will need to levy an additional millage to supplement the foundation program. As much as 15 mills may be levied for this purpose without exceeding the maximum levy permitted without a special vote, but the proceeds may not exceed 25 percent of the amount of the foundation program. Since it can be assumed that the foundation program is not likely to be adequate in the judgment of an affluent community, some additional levy is likely. We assume that an additional levy within the permitted range, equal to 5 mills, will suffice, and hence assume a 45-mill levy against property values determined by the gross proceeds method.

Apart from the proceeds of the severance tax that are earmarked for areas affected by coal development, there are no state laws regulating the rate at which local units of government may tax property within their jurisdiction, nor any equalization mechanism to compensate for local effects of large new investments, other than federal projects.[6] This being the case, current county rates could persist, particularly if the demand for community services for which the county is customarily responsible increases proportionally with the increase in coal extraction and related activities. We examine these questions in detail in chapter 8. However, for a preliminary general overview, we have initially selected a range of between 27 and 36 mills per dollar of taxable property value as the rate for county tax purposes. With this range, and the added 45 mills assumed to prevail for the school district in which the coal extraction and possibly

---

[6] Federal impact funds are typically associated with areas affected by either federal resource development projects or military installations which result in public service demands that exceed the capacity of local units of government to supply them. Federal aid for highways and roads may be one source of impact mitigation that would not distinguish between private or federal sources of the impacting activity.

Table 6-2a. Estimated Big Horn County Gross Proceeds Tax Yields for Local Government Purposes, 1975–90

($000)

Year	Estimated taxable property	Levies		
		72 mills	76 mills	81 mills
Scenario I				
1975–90	25,762	1,855	1,958	2,087
Scenario II				
1975	25,762	1,855	1,958	2,087
1976	31,725	2,284	2,411	2,570
1977	36,225	2,608	2,753	2,934
1978	44,325	3,191	3,369	3,590
1979	38,925	2,803	2,958	3,153
1980–90	42,975	3,094	3,266	3,481
Scenario III				
1975	25,762	1,855	1,958	2,087
1976	31,725	2,284	2,411	2,570
1977	36,427	2,623	2,768	2,951
1978	47,925	3,451	3,642	3,882
1979	50,152	3,611	3,812	4,062
1980	60,975	4,390	4,634	4,939
1981	76,500	5,508	5,814	6,197
1982	79,875	5,751	6,071	6,470
1983–90	87,975	6,334	6,686	7,126
Scenario III (minus 25%)				
1975	25,762	1,855	1,958	2,087
1976	31,725	2,284	2,411	2,570
1977	36,377	2,619	2,765	2,947
1978	47,025	3,386	3,574	3,809
1979	47,346	3,409	3,598	3,835
1980	56,475	4,066	4,292	4,574
1981	68,119	4,905	5,177	5,518
1982	70,650	5,087	5,369	5,723
1983–90	76,725	5,524	5,831	6,215
Scenario III (plus 25%)				
1975	25,672	1,848	1,951	2,079
1976	31,725	2,284	2,411	2,570
1977	36,478	2,626	2,772	2,955
1978	48,825	3,515	3,711	3,955
1979	52,959	3,813	4,025	4,290
1980	65,475	4,714	4,976	5,303
1981	84,881	6,111	6,451	6,875
1982	89,100	6,415	6,772	7,217
1983–90	99,225	7,144	7,541	8,037

Table 6-2b. Estimated Rosebud County Gross Proceeds Tax Yields for Local Government Purposes, 1975–90
($000)

Year	Estimated taxable property	Levies		
		72 mills	76 mills	81 mills
		Scenario I		
1975–90	10,647	767	809	862
		Scenario II		
1975	10,647	767	809	862
1976	18,064	1,301	1,373	1,463
1977	22,444	1,616	1,706	1,818
1978	24,441	1,760	1,858	1,980
1979	28,435	2,047	2,161	2,303
1980–90	30,432	2,191	2,313	2,465
		Scenario III		
1975	10,647	767	809	862
1976	18,267	1,315	1,388	1,480
1977	26,246	1,890	1,995	2,126
1978	39,268	2,827	2,984	3,181
1979	57,460	4,137	4,367	4,654
1980	78,560	5,656	5,971	6,363
1981	93,432	6,727	7,101	7,568
1982	108,957	7,845	8,281	8,826
1983	112,332	8,088	8,537	9,099
1984–90	120,432	8,671	9,153	9,755
		Scenario III (minus 25%)		
1975	10,647	767	809	862
1976	18,216	1,312	1,384	1,475
1977	25,296	1,821	1,923	2,049
1978	35,561	2,560	2,703	2,880
1979	50,204	3,615	3,816	4,067
1980	66,528	4,790	5,056	5,389
1981	77,682	5,593	5,904	6,292
1982	89,326	6,431	6,789	7,235
1983	91,857	6,614	6,981	7,440
1984–90	97,932	7,051	7,443	7,932
		Scenario III (plus 25%)		
1975	10,647	767	809	862
1976	18,318	1,319	1,392	1,484
1977	27,197	1,958	2,067	2,203
1978	42,975	3,094	3,266	3,481
1979	64,716	4,660	4,918	5,242
1980	90,591	6,523	6,885	7,338
1981	109,182	7,861	8,298	8,844
1982	128,588	9,258	9,773	10,416
1983	132,807	9,562	10,093	10,757
1984–90	142,932	10,291	10,863	11,577

related conversion facilities are located, we have combined levies of 72, 76, and 81 mills. Tables 6-2a and 6-2b give the estimated tax base and the gross proceeds tax yields good to a first approximation for Big Horn and Rosebud counties respectively.[7]

DIRECT TAXES ON VALUE OF COAL MINING AND ENERGY CONVERTING FACILITIES

Industrial Property Tax

The taxable value of mines is prescribed by law to be 45 percent of the annual value of mine output. There are also capital outlays for facilities above ground, however, that represent taxable industrial property associated with mining. There is a different method prescribed for computing taxable value for that portion of the investment in mine property. Briefly stated, the taxable value of the investment "above ground" is about 7½ percent of the original investment in facilities. Typically, the assessed value is taken at 40 percent of the "market value," which is taken at 75 percent of actual investment. Most industrial property would fall in that class which has a taxable value equivalent to 30 percent of assessed value. The product of these fractional amounts (0.75 × 0.40 × 0.30) is 9 percent of investment. However, a substantial portion of the investment in a truck and shovel operation is reflected in the large coal carrier vehicles which could be interpreted as falling in the class of vehicles and trucks that has taxable values stipulated to be 20 percent of assessed value. Similarly for the investment in energy conversion facilities, the taxable value of air pollution abatement facilities is only 7 percent of the assessed value, and since as noted earlier this represents roughly 20 percent of the investment, the weighted average taxable value for power plants (Colstrip 1 and 2) approximates 25 percent of the assessed value here as well. When these considerations are accounted for, the taxable value of industrial facilities of the sort we consider for Big Horn and Rosebud counties is roughly 7.5 percent of the original investment. We show both the taxable values and the industrial property tax yields for each scenario at the various levies in tables 6-3a and 6-3b.

---

[7] A more detailed analysis of the determination of tax rates and probable yields will be undertaken in a different context in chapter 8.

Table 6-3a. Estimated Big Horn County Industrial Property Tax Yields, 1975–90

($000)

Year	Estimated taxable industrial property	Levies		
		72 mills	76 mills	81 mills
		Scenario I		
1975–90	3,476	250	264	282
		Scenario II		
1975	3,476	250	264	282
1976	4,229	304	321	343
1977	4,938	356	375	400
1978–90	5,960	429	453	483
		Scenario III		
1975	3,476	250	264	282
1976	4,229	304	321	343
1977	5,028	362	382	407
1978	7,482	539	569	606
1979	9,312	670	708	754
1980	11,307	814	859	916
1981	13,145	946	999	1,065
1982	14,135	1,018	1,074	1,145
1983	14,232	1,025	1,082	1,153
1984–90	14,660	1,056	1,114	1,187
		Scenario IV		
1975	3,476	250	264	282
1976	4,229	304	321	343
1977	5,028	362	382	407
1978	7,482	539	569	606
1979	9,312	670	708	754
1980	11,307	814	859	916
1981	13,145	946	999	1,065
1982	14,135	1,018	1,074	1,145
1983	14,232	1,025	1,082	1,153
1984	14,660	1,056	1,114	1,187
1985–87	23,629	1,701	1,796	1,914
1988–90	53,098	3,823	4,035	4,301
		Scenario IVa		
1975	3,476	250	264	282
1976	4,229	304	321	343
1977	5,028	362	382	407
1978	7,482	539	569	606
1979	9,312	670	708	754
1980	11,307	814	859	916
1981	13,145	946	999	1,065
1982	14,135	1,018	1,074	1,145
1983	14,232	1,025	1,082	1,153
1984	14,660	1,056	1,114	1,187
1985–87	23,629	1,701	1,796	1,914
1988–90	53,098	3,823	4,035	4,301

Table 6-3b. Estimated Rosebud County Industrial Property Tax Yields, 1975–90

($000)

Year	Estimated taxable industrial property	Levies 72 mills	76 mills	81 mills
		Scenario I		
1975	2,336	168	178	189
1976–90	18,806	1,354	1,429	1,523
		Scenario II		
1975	2,336	168	178	189
1976	20,281	1,460	1,541	1,643
1977	21,153	1,523	1,608	1,713
1978	21,470	1,546	1,632	1,739
1979	22,345	1,609	1,698	1,810
1980–90	22,742	1,637	1,728	1,842
		Scenario III		
1975	2,336	168	178	189
1976	20,371	1,467	1,548	1,650
1977	22,765	1,639	1,730	1,844
1978	26,345	1,899	2,002	2,134
1979	30,955	2,229	2,353	2,507
1980	33,842	2,437	2,572	2,741
1981	36,362	2,618	2,764	2,945
1982	38,627	2,781	2,936	3,129
1983	39,617	2,852	3,011	3,209
1984	39,714	2,859	3,018	3,217
1985–90	40,142	2,890	3,051	3,252
		Scenario IV		
1975	2,336	168	178	189
1976	20,371	1,467	1,548	1,650
1977	22,765	1,639	1,730	1,844
1978	26,345	1,897	2,002	2,134
1979	30,955	2,229	2,353	2,507
1980	33,842	2,437	2,572	2,741
1981	36,362	2,618	2,764	2,945
1982	120,059	8,644	9,124	9,725
1983	121,049	8,716	9,200	9,805
1984	121,146	8,723	9,207	9,813
1985–90	203,006	14,616	15,428	16,443
		Scenario IVa		
1975	2,336	168	178	189
1976	20,371	1,467	1,548	1,650
1977	22,765	1,639	1,730	1,844
1978	26,345	1,897	2,002	2,134
1979	30,955	2,229	2,353	2,507
1980	33,842	2,437	2,572	2,741
1981	36,362	2,618	2,764	2,945
1982	120,059	8,644	9,124	9,725
1983	121,049	8,716	9,200	9,805
1984	121,146	8,723	9,207	9,813
1985–87	121,574	8,753	9,240	9,847
1988–90	203,006	14,616	15,428	16,443

TAX REVENUE ASSOCIATED INDIRECTLY WITH COAL AND COAL-RELATED ACTIVITIES

Real and Personal Property Taxes

Developments in the coal fields and related economic activities, whether associated with mining for export only or for energy conversion at site, will induce some expansion in population, along with employment. New additions to the population will bring with them, and also acquire, personal property as well as real estate, all of which is taxable. Indeed, this source of taxable property will be the main source of tax revenue available to some units of local government. While this source of revenue is not large in relation to other coal-related revenue sources, it may be the principal tax source for towns and cities not having access to one or more large industrial properties.

Important as the real and personal property tax base is in estimating the capacity of some localities to meet community services, we know very little about the composition of real and personal property assets of new immigrants responding to coal developments in eastern Montana. One might speculate about the behavior of new additions to the work force and population and build up in simulated fashion the expected acquisition of real and personal property. An implication of such an exercise might be that the real property values, increasing in response to the increase in demand for real estate, and rising construction costs, would produce taxable values considerably above current values per capita. Polzin has attempted to construct real and personal property tax profiles of new entrants, and while the unpublished study represents a more or less informal analysis, he has concluded nonetheless that the taxable value per capita does not differ significantly from the per capita taxable value of real and personal property on the tax rolls prior to development.[8] It may well be that the composition of assets owned by members of the two subpopulations is different. That is, there is likely to be a difference between the amount and kinds of land held as well as the composition of the personal property. We might expect a disproportionate representation of ranches and livestock among the holdings of established property owners compared with the more recent additions to the population. In any event, even with higher costs for real estate and residential dwellings, the very different composition of taxable assets could result in equivalent per capita taxable property values for the two groups.

[8] Conveyed to the authors in personal communication.

Results of our limited and admittedly inadequate analysis suggest nothing that would be inconsistent with this. Noting that real property has not been reappraised since 1969 in Rosebud County, any changes in the value of taxable property reflect additions to the tax rolls rather than revaluation of property existing on the rolls prior to development in the coal fields. Annual changes in the taxable value of real and personal property for school districts for which we have such data allow us to determine the relationship between the value of additional property in such districts and changes in population.

School district 4, in which Forsyth is located, and which serves a substantial part of the population associated with construction at Colstrip 1 and 2, along with the employment at the Western Energy mine, does not itself have any large industrial property. The increase in the value of taxable property in school district 4, then, is attributable to the non-industrial real and personal property associated with the migrants domiciled in Forsyth and environs. Data on the changing amount of taxable value are readily obtainable.[9]

Data on the change in population are available, but since these data are not household enumerations of the sort provided by the decennial census, the estimated changes in population are not as reliable as the tax data taken off the rolls. Intracensual year estimates are built up from data on school enrollment and automobile registrations and cannot be checked for accuracy without an actual enumeration. For example, while we have the data on number of students enrolled each year by school district, and we have the relation between students and total population for the census year, we do not know whether that ratio will hold if the composition of the population changes because of the influx of construction workers, and other perhaps more subtle changes in population characteristics. Inadequate as these estimates may be, we have "constructed" the population changes from student enrollment data assuming the fraction of total population accounted for by students to be 0.25 as in the census year, and have developed from this an estimate of $2,000 (+ 25 percent) as a crude approximation of the per capita taxable value of real and personal property. This tends to be a bit lower than Polzin's estimate (1974) and thus represents a conservative estimate which may be warranted, owing to the uncertainty surrounding such an approximation.

---

[9] These have been provided us by courtesy of John Clark, Division of Research, Montana Department of Revenue.

With these data representing our estimate of the per capita taxable real and personal property, and our projected population for each scenario, we can develop estimates of the increment in population for each county and the estimated increment in taxable value of real and personal property as shown in tables 6-4a and 6-4b. The revenue yield from this source, of course, will depend on what rates are levied against this component of taxable property. Considering the relatively small communities in Montana, we find that a range of between 48 and 66 mills includes the majority of the tax rates levied by such communities. Similarly, except for school districts that have large industrial properties, and where something like 45 mills would be relevant, the rates range upward to 120 mills. For our purpose then, we take 45 mills as the minimum, 82 mills as an intermediate, 120 mills as a high estimate. Combining all three (county, school, and city) levies, we have a rate of 120 mills for the minimum, an estimate of 170 mills intermediate in the range, and a high of 222 mills. These are shown in tables 6-4a and 6-4b.

It might be noted that different rates could be assumed that may be more or less relevant to a particular community or case, but those shown in tables 6-4a and b are good to a first approximation for rough comparisons.[10] Secondly, it should be noted that the portion of the real and personal property that does not fall within a town or city taxing jurisdiction would be taxed at a rate of from 48 to 66 mills less; that is, only the county and school district levies would be relevant. The yields with the town levies included are shown, however, to permit examining the fiscal circumstances of cities and towns.

SUMMARY

In this chapter we have identified and sought to obtain a preliminary overview of the sources and amounts of tax revenues that would be available to local units of government where the impact of coal development would be felt. The relevant portion of the yield from the state severance tax was separated out for subsequent comparison with the implied expenditures to service community needs. The severance tax yields that are earmarked for use of local governments are very large compared with any other single source of revenue available at the local level.

---

[10] More detailed analysis in a different context is deferred until chapter 8.

Table 6-4a. Increment to Real and Personal Taxable Property Associated with Big Horn County Coal Development
($000)

Year	Projected incremental population	Projected incremental taxable property	Local government receipts Levies		
			120 mills	170 mills	222 mills
		Scenario I			
1975	3,047	6,094	731	1,036	1,353
1976	2,807	5,614	674	954	1,246
1977	2,676	5,352	642	910	1,188
1978	2,649	5,298	636	901	1,176
1979	2,533	5,066	608	861	1,125
1980	2,630	5,260	631	894	1,168
1981	2,456	4,912	589	835	1,090
1982	2,586	5,172	621	879	1,148
1983	2,429	4,858	583	826	1,078
1984	2,512	5,024	603	854	1,115
1985	2,333	4,666	560	793	1,036
1986	2,427	4,854	582	825	1,078
1987	2,456	4,912	589	835	1,090
1988	2,324	4,648	558	790	1,032
1989	2,421	4,842	581	823	1,075
1990	2,301	4,602	552	782	1,022
		Scenario II			
1975	3,047	6,094	731	1,036	1,353
1976	3,256	6,512	781	1,107	1,446
1977	3,567	7,134	856	1,213	1,584
1978	4,336	8,672	1,041	1,474	1,925
1979	4,017	8,034	964	1,366	1,784
1980	4,034	8,068	968	1,372	1,791
1981	4,180	8,360	1,003	1,421	1,856
1982	4,320	8,640	1,037	1,469	1,918
1983	4,133	8,266	992	1,405	1,835
1984	4,235	8,470	1,016	1,440	1,880
1985	4,031	8,062	967	1,371	1,790
1986	4,148	8,296	996	1,410	1,842
1987	3,998	7,996	960	1,359	1,775
1988	4,116	8,232	988	1,399	1,028
1989	3,976	7,952	954	1,352	1,765
1990	4,104	8,208	985	1,395	1,822
		Scenario III			
1975	3,047	6,094	731	1,036	1,353
1976	3,360	6,720	806	1,142	1,492
1977	4,292	8,584	1,030	1,459	1,906
1978	5,150	10,300	1,236	1,751	2,287
1979	5,631	11,262	1,351	1,915	2,500
1980	6,361	12,722	1,527	2,163	2,824
1981	6,636	13,272	1,593	2,256	2,946
1982	6,703	13,406	1,609	2,279	2,976
1983	6,728	13,456	1,615	2,288	2,987
1984	6,810	13,620	1,634	2,315	3,024

Table 6-4a (Continued)

($000)

Year	Projected incremental population	Projected incremental taxable property	Local government receipts Levies		
			120 mills	170 mills	222 mills
1985	6,887	13,774	1,653	2,342	3,058
1986	6,721	13,442	1,613	2,285	2,984
1987	6,867	13,734	1,648	2,335	3,050
1988	6,708	13,416	1,610	2,281	2,978
1989	6,861	13,722	1,647	2,333	3,046
1990	6,717	13,434	1,612	2,284	2,982
		Scenario IV			
1975	3,047	6,094	731	1,036	1,353
1976	3,360	6,720	806	1,142	1,492
1977	4,292	8,584	1,030	1,459	1,906
1978	5,149	10,298	1,236	1,751	2,286
1979	5,630	11,260	1,351	1,914	2,500
1980	6,361	12,722	1,527	2,163	2,824
1981	7,429	14,858	1,783	2,525	3,298
1982	8,527	17,054	2,046	2,899	3,786
1983	11,779	23,558	2,827	4,005	5,230
1984	10,436	20,872	2,505	3,548	4,634
1985	10,940	21,880	2,626	3,720	4,857
1986	11,117	22,234	2,668	3,780	4,936
1987	10,906	21,812	2,617	3,708	4,842
1988	11,089	22,178	2,661	3,770	4,924
1989	10,924	21,848	2,622	3,714	4,850
1990	11,131	22,262	2,671	3,785	4,942
		Scenario IVa			
1975	3,047	6,094	731	1,036	1,353
1976	3,360	6,720	806	1,142	1,492
1977	4,292	8,584	1,030	1,459	1,906
1978	5,149	10,298	1,236	1,751	2,286
1979	5,630	11,260	1,351	1,914	2,500
1980	6,361	12,722	1,527	2,163	2,824
1981	7,430	14,860	1,783	2,526	3,299
1982	8,528	17,056	2,047	2,900	3,786
1983	11,781	23,562	2,827	4,006	5,231
1984	10,437	20,874	2,505	3,549	4,634
1985	10,938	21,876	2,625	3,719	4,856
1986	11,115	22,230	2,668	3,779	4,935
1987	10,905	21,810	2,617	3,708	4,842
1988	11,088	22,176	2,661	3,770	4,923
1989	10,923	21,846	2,622	3,714	4,850
1990	11,130	22,260	2,671	3,784	4,942

Table 6-4b. Increment to Real and Personal Taxable Property Associated with Rosebud County Coal Development

($000)

Year	Projected incremental population	Projected incremental taxable property	Local government receipts Levies		
			120 mills	170 mills	222 mills
		Scenario I			
1975	4,666	9,332	1,120	1,586	2,072
1976	3,322	6,644	797	1,129	1,475
1977	2,239	4,478	537	761	994
1978	2,037	4,074	489	693	904
1979	1,929	3,858	463	656	856
1980	2,000	4,000	480	680	888
1981	1,966	3,932	471	668	873
1982	2,096	4,192	503	713	931
1983	2,173	4,346	522	739	965
1984	2,383	4,766	572	810	1,058
1985	2,499	4,998	600	850	1,110
1986	2,738	5,476	657	931	1,216
1987	2,899	5,798	696	986	1,287
1988	3,184	6,368	764	1,083	1,414
1989	3,388	6,776	813	1,152	1,504
1990	3,732	7,464	896	1,269	1,660
		Scenario II			
1975	4,666	9,332	1,120	1,586	2,072
1976	4,021	8,042	965	1,367	1,785
1977	3,352	6,704	804	1,140	1,488
1978	3,254	6,508	781	1,106	1,445
1979	3,509	7,018	842	1,193	1,558
1980	3,850	7,700	924	1,309	1,709
1981	3,901	7,802	936	1,326	1,732
1982	4,081	8,162	979	1,388	1,812
1983	4,144	8,288	995	1,409	1,840
1984	4,366	8,732	1,048	1,484	1,939
1985	4,523	9,046	1,086	1,538	2,008
1986	4,800	9,600	1,152	1,632	2,131
1987	4,966	9,932	1,192	1,688	2,205
1988	5,277	10,554	1,266	1,794	2,343
1989	5,492	10,984	1,318	1,867	2,438
1990	5,863	11,726	1,407	1,993	2,603
		Scenario III			
1975	4,937	9,874	1,185	1,679	2,192
1976	5,282	10,564	1,268	1,796	2,345
1977	4,717	9,434	1,132	1,604	2,094
1978	5,129	10,258	1,231	1,744	2,277
1979	5,917	11,834	1,420	2,012	2,627
1980	6,760	13,520	1,622	2,298	3,001
1981	7,374	14,748	1,770	2,507	3,274
1982	7,802	15,604	1,872	2,653	3,464
1983	7,958	15,916	1,910	2,706	3,533
1984	8,202	16,404	1,968	2,789	3,642

Table 6-4b (Continued)

($000)

Year	Projected incremental population	Projected incremental taxable property	Local government receipts Levies		
			120 mills	170 mills	222 mills
1985	8,427	16,854	2,022	2,865	3,742
1986	8,559	17,118	2,054	2,910	3,800
1987	8,891	17,782	2,134	3,023	3,948
1988	9,112	18,224	2,187	3,098	4,046
1989	9,520	19,040	2,285	3,237	4,227
1990	9,810	19,620	2,354	3,335	4,356
		Scenario IV			
1975	4,937	9,874	1,185	1,679	2,192
1976	5,532	11,064	1,328	1,881	2,456
1977	6,208	12,416	1,490	2,111	2,756
1978	9,293	18,586	2,230	3,160	4,126
1979	12,879	25,758	3,091	4,379	5,718
1980	15,556	31,112	3,733	5,289	6,907
1981	16,408	32,816	3,938	5,579	7,285
1982	17,479	34,958	4,195	5,943	7,761
1983	16,724	33,448	4,014	5,686	7,425
1984	14,235	28,470	3,416	4,840	6,320
1985	12,779	25,558	3,067	4,345	5,674
1986	12,069	24,138	2,897	4,103	5,359
1987	11,936	23,872	2,865	4,058	5,300
1988	11,949	23,898	2,868	4,063	5,305
1989	11,888	23,776	2,853	4,042	5,278
1990	12,199	24,398	2,928	4,148	5,416
		Scenario IVa			
1975	4,937	9,874	1,185	1,679	2,192
1976	5,532	11,064	1,328	1,881	2,456
1977	6,208	12,416	1,490	2,111	2,756
1978	9,293	18,586	2,230	3,160	4,126
1979	12,667	25,334	3,040	4,307	5,624
1980	14,198	28,396	3,408	4,827	6,304
1981	12,065	24,130	2,896	4,102	5,357
1982	11,072	22,144	2,657	3,764	4,916
1983	11,329	22,658	2,719	3,852	5,030
1984	13,909	27,818	3,338	4,729	6,176
1985	16,643	33,286	3,994	5,659	7,389
1986	16,908	33,816	4,058	5,749	7,507
1987	14,497	28,994	3,479	4,929	6,437
1988	13,375	26,750	3,210	4,547	5,939
1989	12,639	25,278	3,033	4,297	5,612
1990	12,651	25,302	3,036	4,301	5,617

Indeed, the county and local governments' allotment from estimated severance tax receipts could be greater than the combined revenues from mine (gross proceeds), industrial, and real and personal property taxes.[11]

Of the three sources of revenue from property taxation, the nonindustrial real and personal property tax yield is generally the smallest. Those cities and towns, and corresponding school districts, that do not have mines and industrial plants in their tax bases will be disadvantaged when it comes to financing public services. But it should be noted that funds from the severance tax earmarked for impact areas are substantially greater than the combined yields from the gross proceeds and industrial property taxes. Accordingly, failure to have such facilities located within its taxing jurisdiction need not spell fiscal disaster for a local community, provided that the Coal Board is responsive to its needs. The relative importance of these considerations will be taken up systematically in the next two chapters.

One overall qualification is in order. The estimates presented in this and the previous chapter are good only to a first approximation for those who may be interested in the relative distribution of revenues among the source, the distribution between the state and local entities of government, and among the various scenarios, given the characteristics that we postulated for each. Throughout we have sought to communicate a sense of the approximate nature of the estimates. Aside from the matter of departures in reality from the assumptions we have made, there are, as we have noted, discrepancies caused by divergences between model outputs and what we have reason to feel is likely to happen.

One matter we have not mentioned, though it is not significant, may nonetheless be worth noting to indicate our awareness of it. This has to do with the operation of the mandated minimum school levies, and the availability of the yields from one school district for use by the county to make up a deficiency in other districts within the same county. This is reflected in our estimates; however, to anticipate the results of chapter 7, we did not carry out a further refinement that would affect the distribution of yields between the state and the local governments. That relates to the use of the excess being restricted to financing only *foundation* programs

---

[11] Funds earmarked for the local impact account are not required to be disbursed to the county where they originated. Indeed, it is not likely that all such funds originating from severance taxes on Big Horn and Rosebud county coal will return to these counties.

in other school districts within the county. Any surplus above these needs is destined for the *state equalization account.* If there is a surplus of this sort in the two counties, as there is likely to be, we have tended to understate by a small amount the share of total tax revenues going to the state and overstate by a small amount those remaining with the county.

Because the financing of educational services is a significant portion of the financing of local public services, and because of the complexities involved, we address some of these issues in greater detail in the following chapter.

7 Community Services and Expenditures—Educational

INTRODUCTION

The extensive and rapid expansion of coal extraction and possible energy conversion will be accompanied by a rapid growth in population, as we have noted in chapter 3. Depending on the scenario and county, population could increase from 25 percent between 1975 and 1985 as a minimum for scenario I, Big Horn County, to more than double as in the case of scenario IV for Rosebud County. Moreover, the peak employment and population would be experienced during the peak construction period when the population, under the most extreme conditions postulated, could more than treble in Rosebud County. A significant proportion of the additions to the population would be school-aged children who would require educational facilities and services greatly exceeding those currently available. Indeed, provision of educational services for the expanded population would represent the major single public expenditure category. Currently only about half of the tax-financed public expenditures at the county level are accounted for by education, whereas about three-quarters of the total tax-financed expenditures at the city and town level represent outlays for education. Since educational expenditures bulk so large in the total budget for local government services, we have elected to treat this public service expenditure category separately.

We first address the question of estimating school enrollment and the costs per student given in constant (1974) dollars in order to express

104

both expenditures and receipts in comparably valued monetary units. Following the description of our method of estimating school enrollment and annual expenditures, we shall present the estimated magnitudes for each county separately for each scenario.

ESTIMATED ENROLLMENT AND COSTS PER STUDENT

School enrollment is derived from estimates of the population. Population estimates associated with a given change in economic activity (as those in chapter 3) are good only to a first approximation. When the estimates need to be made for relatively small geographic areas, accuracy is further diminished. Indeed, population estimates of local areas for intracensual years are subject to a wide margin of error if perturbations to the local economy are anything like those postulated in our large expansion scenarios for eastern Montana. Typically, population estimates for intracensual years are built up from a combination of data, recorded school enrollment and automobile registrations being chief among these. The results of these methods may exhibit a wide margin of error because temporary (week day) residence may be established at the construction site by workers whose families remain housed in other localities. There is also seasonality in construction employment in eastern Montana, with some construction workers returning to homes in urban areas like Billings or Miles City for the winter months. The relation of local employment to local population during the construction phase of new developments also may be different from that between the operating work force and the local population once construction is completed and a more permanent body of operating labor takes over the principal functions in the local labor force. A full understanding of the changes that take place in the labor force and the relation between employment and population, along with the age distribution of the population during the period it is taking place, would require what are commonly referred to as "longitudinal" studies; that is, continuous monitoring of conditions on which we wish to inform ourselves over the entire period from the initial perturbation to the reestablishment of a steady-state community structure.

Longitudinal studies are frequently recommended, almost always desired, but almost never undertaken. To our knowledge, there are none on examples of "boom towns" of the sort relevant to our problem. Since there are no ready references that yield useful information, and the purpose of this analysis is to evaluate impacts of postulated options in advance of the decisions that would implement one or another of them, we must rely on less precise data, along with sensitivity analysis, to set the range of variation that could result from errors in assumptions or data.

Our county population projections are obtained from the econometric model described in chapter 3. There is no explicit, built-in allowance for the possibility that construction labor employed in Rosebud County might be housed in part, say, in Miles City (Custer County), nor that both construction and operating labor in the Decker area of Big Horn County might be commuting from residences in Sheridan, Wyoming. In both cases, inferring a school-age population and school enrollment from the population based on growth of employment associated with activities in each of our scenarios will overstate the demand for educational services. However, we believe it is preferable to demarcate an upper bound in this fashion rather than risk underestimating required expenditures in an effort to forecast school enrollment more precisely.

Given the annual increase in population associated with each of the possible scenarios we have postulated, we can infer the implicit school-age population corresponding to it. That is, the ratio of total population to school-age population is about 4 to 1. This holds true for North Dakota[1] and for Montana state-wide. It is not clear, however, that the new population will have an age distribution identical to that of the original population. On the one hand we can speculate that construction crews have an overrepresentation of single men—or that place of residence is in a county different from the county of employment, so that a smaller proportion of the projected population will in fact appear as students enrolled in the county of employment. Alternatively, one can speculate that the younger cohorts of the construction labor force, those with a higher likelihood of young school-age children, are the most likely to respond to construction opportunities involving relocation of households. Accordingly, a higher ratio of students to total population might be postulated as a possibility.[2] Without an actual census of the new population via longitudinal monitoring, there is no way of knowing with certainty which of the outcomes is the more likely. For our purpose we begin with the assumption that the ratio of school-age children to the total population associated with the increment in employment and economic

[1] Arlen G. Leholm, Larry F. Leistritz, and Thor A. Hertsgaard, *Local Impacts of Energy Resource Development in the Northern Great Plains.*

[2] This outcome is thought to be less likely by some who have investigated these phenomena in similar situations. We owe this point to Jack Schanz of Resources for the Future.

activities induced by coal and related developments is the same as for the resident population. This would give us an increment in the school-age population equivalent to a quarter of the increment in the population at large in the county in question. To determine finally how sensitive our results will be to the precise ratio assumed, we make alternative assumptions of 30 and 20 percent of the increment in the population at large as representing the school-age population.

We further assume that the distribution of the students between elementary and high schools is similar to the distribution given in the 1970 decennial census. The enrollment in elementary school accounted for about 76 percent of the total enrollment in both counties, with the remainder in high school. It is not clear, however, that this ratio will continue, as it appears to be a reflection of the more heavy attrition between elementary and high school enrollment in rural compared with more urban areas. If there were no attrition, and assuming eight grades in elementary school and four in high school, the distribution would be exactly two-thirds and one-third. There is some evidence that this is approached in urban areas, and thus in order to test the ultimate results for sensitivity to these changes, we also assume alternatively a 70-30 split for comparison.

Given the increase in population derived from the model described in chapter 3, the split between school-age population and population at large, and between elementary and high school enrollment, we assume that 19 percent of the increment in population will be enrolled in elementary schools, while 6 percent will be enrolled in high schools. (These percentages are altered to test for sensitivity, as indicated above.)

Once the increased enrollment in elementary or high schools is estimated in this manner for each scenario, estimates are made of the total increase in investment in facilities required by this enrollment increase, the annualized costs, and the operating costs, all in 1974 dollars. Investment in facilities is based on a standard 105 square feet per elementary school student and 120 square feet per high school student, with costs estimated at $40 per square foot for school construction. Investment is indicated when an increase in enrollment exceeds any previous peak enrollment; that is, since there will likely be a surge in enrollment during the construction period, some excess capacity may exist for a time following a construction project until the enrollment may again exceed the previous peak, requiring additional investment. While it is possible that any surge in enrollment associated with a temporary construction-related population increase may be met by bringing in portable classrooms, thus avoiding a permanent investment in what might later be excess capacity, we have not attempted to introduce this refinement into the analysis. The investment is converted into annualized costs by assuming a range of interest rates and amortization terms. That is, annualized costs range from a low produced by the combination of an interest rate of 6 percent and a thirty-year term, to a high produced by a combination of 8 percent and twenty years. It isn't possible to predict what financing terms would be available, or whether the shorter or longer amortization period is the more realistic. For our purposes we elected to use an interest rate of 7 percent, which is thought to be conservative for bonds having tax-exempt interest earnings, and a twenty-year term that would be shorter than the physical serviceability of the facilities. Annualized costs obtained in this manner were combined with annual operating expenses taken to be $1,014 per student per year, as reflected in Montana experience for the 1974–75 school year. This provides an annual estimate of educational expenditures for the projected school population induced by the coal and related developments.

Tables 7-1a and 7-1b give an example of the computations. Here we have the results assuming 250 students per thousand additional population, a 76-24 elementary-high school split and the projected population associated with scenario IV, for Big Horn County. If we elected to check the difference in the estimated magnitudes, assuming alternatively 300 and 200 students per thousand population, all results would be scaled either up or down by 20 percent. Similarly, if we assume an elementary-high school split closer to the average for the state, namely a 70-30 split, we would anticipate as much as a 4 percent increase in the annualized investment costs owing to the greater space requirements, hence investment, for high school facilities per student.

Some additional comments concerning the table are in order. Years are designated from 1 through 20. Actually the model was backdated to 1971 in order to pick up the developments that had taken place since the last decennial census, both in connection with the coal extraction between 1970 and 1975 and the stepped-up activity in the Colstrip area associated with the construction of the steam electric power plant units (Colstrip 1 and 2). Coal production in Rosebud County increased from about 3 million tons in 1970 to a reported 6 million in 1973 and an estimated 8 million by 1975.[3] Similarly, the development in Big Horn

---

[3] Reports of the Montana State Department of Revenue for the periods July 1, 1970 to June 30, 1972; June 1, 1972 to June 30, 1974 and "Coal Development Information Packet," Montana Energy Advisory Council, December 1974.

County went from less than 100,000 tons reported in 1970 to a reported 4 million in 1973 and an estimated 12 million by 1975. Data for the decade 1975–85 accordingly begin with the fifth and continue through the fourteenth entry.

Next a caveat and qualification. Despite the expansion of coal mined between 1970 and 1975 in Big Horn County, there is little documentable evidence that there has been an expansion of population corresponding to the expansion of employment. Neither school district reports to the state equalization board nor independently sought estimates of population growth[4] indicate anything like the population expansion that should have attended the increase in mining activity. Speculation has it that school enrollment does not reflect the increase in population because of potentially better educational facilities in Sheridan, Wyoming, and thus only construction and operating laborers without school-age children locate nearer the site of their work. Since intracensual year estimates by county rely heavily on the number of students reported enrolled in district schools, this conventional basis for estimating small area populations would greatly bias the results. It is not feasible for us to investigate this matter intensively. Nor could we be certain that even if employees working in the area are domiciled predominantly in Sheridan, this behavior would persist for the employment expansion postulated for the higher activity level scenarios like III and IV. And as we have already indicated, for our purposes it is preferable to assume that the population normally associated with a given labor force is domiciled within the county in which the economic activity occurs. If we are concerned about the impact of increased employment and population on community services and fiscal structure of the coal region, it is preferable to address and evaluate an upper bound case rather than risk underestimating the magnitude of the impact. Accordingly, quite apart from the problem we would encounter in attempting to document the population increase between 1970 and our starting point, 1975, from existing sources, we elect to rely on the population estimates generated by the forecasting model described in chapter 3.

With the above qualifications and caveats as background, we present in tables 7-2a and 7-2b the estimated educational expenditures corresponding to each scenario postulated for Big Horn and Rosebud counties.

---

[4] Study prepared for Westmoreland Resources by the consulting firm, Mountain West Research, Inc.

Table 7-1a. School Enrollment and Educational Expenditure Forecasts, Big Horn County, Scenario IV

		Elem. enroll. (above stable)	Gain over previous peak			Annualized costs					
Year	Population increment			New investment	Cumulative investment	6% 20 yr.	6% 30 yr.	7% 20 yr.	7% 30 yr.	8% 20 yr.	8% 30 yr.
					Elementary school						
1971	43	8.	8.	34,314.	34,314.	3,014.	2,493.	3,239.	2,765.	3,495.	3,048.
1972	200	46.	38.	159,600.	193,914.	14,021.	11,595.	15,065.	12,862.	16,255.	14,177.
1973	800	198.	152.	638,400.	832,314.	56,083.	46,380.	60,259.	51,449.	65,021.	56,709.
1974	1,190	424.	226.	949,619.	1,781,932.	83,424.	68,990.	89,635.	76,530.	96,719.	84,355.
1975	814	579.	155.	649,573.	2,431,504.	57,065.	47,191.	61,313.	52,349.	66,159.	57,702.
1976	313	638.	59.	249,774.	2,681,277.	21,943.	18,146.	23,576.	20,129.	25,439.	22,187.
1977	932	815.	177.	743,736.	3,425,013.	65,337.	54,032.	70,201.	59,938.	75,750.	66,066.
1978	857	978.	163.	683,885.	4,108,898.	60,079.	49,684.	64,552.	55,114.	69,654.	60,750.
1979	481	1,070.	91.	383,839.	4,492,736.	33,720.	27,886.	36,231.	30,934.	39,094.	34,096.
1980	731	1,209.	139.	583,338.	5,076,073.	51,246.	42,379.	55,061.	47,011.	59,413.	51,818.
1981	1,068	1,412.	203.	852,264.	5,928,336.	74,871.	61,917.	80,445.	68,684.	86,803.	75,707.
1982	1,098	1,620.	209.	876,204.	6,804,540.	76,975.	63,656.	82,705.	70,613.	89,241.	77,833.
1983	3,252	2,238.	618.	2,595,095.	9,399,635.	227,979.	188,534.	244,951.	209,139.	264,310.	230,522.
1984	−1,343	1,983.	0.	0.	9,399,635.	0.	0.	0.	0.	0.	0.
1985	504	2,079.	0.	0.	9,399,635.	0.	0.	0.	0.	0.	0.
1986	177	2,112.	0.	0.	9,399,635.	0.	0.	0.	0.	0.	0.
1987	−211	2,072.	0.	0.	9,399,635.	0.	0.	0.	0.	0.	0.
1988	183	2,107.	0.	0.	9,399,635.	0.	0.	0.	0.	0.	0.
1989	−165	2,076.	0.	0.	9,399,635.	0.	0.	0.	0.	0.	0.
1990	207	2,115.	0.	0.	9,399,635.	0.	0.	0.	0.	0.	0.

High school

Year											
1971	43	3.	3.	12,384.	12,384.	1,088.	900.	1,169.	998.	1,261.	1,100.
1972	200	15.	12.	57,600.	69,984.	5,060.	4,185.	5,437.	4,642.	5,867.	5,117.
1973	800	63.	48.	230,400.	300,384.	20,241.	16,739.	21,747.	18,568.	23,466.	20,466.
1974	1,190	134.	71.	342,720.	643,104.	30,108.	24,899.	32,349.	27,620.	34,906.	30,444.
1975	814	183.	49.	234,432.	877,536.	20,595.	17,031.	22,128.	18,893.	23,877.	20,825.
1976	313	202.	19.	90,144.	967,680.	7,919.	6,549.	8,509.	7,265.	9,181.	8,007.
1977	932	258.	56.	268,415.	1,236,094.	23,580.	19,500.	25,336.	21,632.	27,338.	23,843.
1978	857	309.	51.	246,817.	1,482,910.	21,683.	17,931.	23,297.	19,891.	25,138.	21,925.
1979	481	338.	29.	138,527.	1,621,437.	12,170.	10,064.	13,076.	11,164.	14,109.	12,305.
1980	731	382.	44.	210,529.	1,831,965.	18,495.	15,295.	19,872.	16,966.	21,442.	18,701.
1981	1,058	446.	64.	307,584.	2,139,549.	27,021.	22,346.	29,033.	24,788.	31,327.	27,323.
1982	1,098	512.	66.	316,223.	2,455,772.	27,780.	22,974.	29,848.	25,484.	32,207.	28,090.
1983	3,252	707.	195.	936,577.	3,392,348.	82,278.	68,042.	88,403.	75,479.	95,390.	83,196.
1984	−1,343	626.	0.	0.	3,392,348.	0.	0.	0.	0.	0.	0.
1985	504	656.	0.	0.	3,392,348.	0.	0.	0.	0.	0.	0.
1986	177	667.	0.	0.	3,392,348.	0.	0.	0.	0.	0.	0.
1987	−211	654.	0.	0.	3,392,348.	0.	0.	0.	0.	0.	0.
1988	183	665.	0.	0.	3,392,348.	0.	0.	0.	0.	0.	0.
1989	−165	655.	0.	0.	3,392,348.	0.	0.	0.	0.	0.	0.
1990	207	668.	0.	0.	3,392,348.	0.	0.	0.	0.	0.	0.

Note: Table assumes 250 students per 1,000 population.

Table 7-1b. Cumulative Annualized Investment Costs and Operating Costs for Elementary and High School, Big Horn County, Scenario IV

Year	Cumulative investment	Cumulative annualized investment costs (7%, 20 yr.)	Operating costs ($1,014/student)	Sum (cum. annual plus operating)
1971	46.698.	4,408.	10,900.	15,308.
1972	263,898.	24,909.	61,600.	86,510.
1973	1,132,697.	106,915.	264,400.	371,316.
1974	2,425,035.	228,899.	566,065.	794,964.
1975	3,309,039.	312,340.	772,414.	1,084,754.
1976	3,648,956.	344,425.	851,760.	1,196,184.
1977	4,661,107.	439,962.	1,088,021.	1,527,982.
1978	5,591,808.	527,811.	1,305,271.	1,833,081.
1979	6,114,173.	577,117.	1,427,204.	2,004,320.
1980	6,908,038.	652,050.	1,612,513.	2,264,562.
1981	8,067,885.	761,528.	1,883,251.	2,644,778.
1982	9,260,312.	874,081.	2,161,594.	3,035,674.
1983	12,791,983.	1,207,434.	2,985,976.	4,193,410.
1984	12,791,983.	1,207,434.	2,645,525.	3,852,959.
1985	12,791,983.	1,207,434.	2,773,289.	3,980,723.
1986	12,791,983.	1,207,434.	2,818,159.	4,025,593.
1987	12,791,983.	1,207,434.	2,764,670.	3,972,104.
1988	12,791,983.	1,207,434.	2,811,061.	4,018,495.
1989	12,791,983.	1,207,434.	2,769,233.	3,976,667.
1990	12,791,983.	1,207,434.	2,821,708.	4,029,142.

Note: Table assumes 250 students per 1,000 population.

Table 7-2a. Estimated Incremental Educational Expenditures, Big Horn County, 1975–90

($000)

Year	Scenario I	Scenario II	Scenario III	Scenario IV	Scenario IVa
1975	1,085	1,085	1,085	1,085	1,085
1976	1,024	1,159	1,196	1,196	1,196
1977	991	1,302	1,528	1,528	1,528
1978	984	1,544	1,833	1,833	1,833
1979	954	1,463	2,005	2,004	2,004
1980	979	1,467	2,265	2,265	2,265
1981	935	1,504	2,362	2,645	2,645
1982	968	1,540	2,386	3,036	3,036
1983	928	1,492	2,395	4,193	4,194
1984	949	1,518	2,424	3,853	3,853
1985	904	1,466	2,452	3,981	3,980
1986	928	1,496	2,410	4,026	4,025
1987	935	1,458	2,447	3,972	3,972
1988	901	1,488	2,406	4,018	4,018
1989	926	1,452	2,445	3,977	3,977
1990	896	1,485	2,409	4,029	4,029

Note: Table assumes 250 students per 1,000 population.

Table 7-2b. Estimated Incremental Educational Expenditures, Rosebud County, 1975–90

($000)

Year	Scenario I	Scenario II	Scenario III	Scenario IV	Scenario IVa
1975	1,662	1,662	1,758	1,758	1,758
1976	1,321	1,498	1,881	1,970	1,970
1977	1,047	1,329	1,738	2,211	2,211
1978	995	1,304	1,842	3,310	3,310
1979	968	1,368	2,107	4,587	4,511
1980	986	1,455	2,408	5,540	5,057
1981	977	1,468	2,626	5,844	4,516
1982	1,010	1,513	2,779	6,225	4,264
1983	1,030	1,529	2,834	6,034	4,329
1984	1,083	1,586	2,921	5,403	4,983
1985	1,112	1,626	3,001	5,034	5,927
1986	1,173	1,710	3,048	4,854	6,022
1987	1,214	1,769	3,167	4,820	5,411
1988	1,286	1,879	3,245	4,823	5,126
1989	1,338	1,956	3,391	4,808	4,940
1990	1,425	2,088	3,494	4,887	4,943

Note: Table assumes 250 students per 1,000 population.

RELATION BETWEEN EDUCATIONAL EXPENDITURES AND RESOURCES

Scenario I: Big Horn

In figure 7-1a we display all of the information on revenues from different sources and assumed circumstances that were discussed in chapter 6, along with the expenditures presented in the preceding section that are relevant to Big Horn County for scenario I. As previously mentioned, the enrollment corresponds to the estimated number of students that would be associated with a population domiciled in the county in which its members were employed. This will most likely result in an overestimate of school expenditures. Similarly, the estimated increment in personal income taxes earmarked for state school equalization also would be overestimated, and hence would be less available for local educational support.[5] But this would be of significance to the local school system only in

---

[5] Since the state of Montana would not receive state personal income taxes on residents in Wyoming regardless of the income being earned in Montana, an estimated annual state income tax of $1 to $1.25 million corresponding to the earnings associated with scenario I would be unavailable for supporting the state school equalization fund.

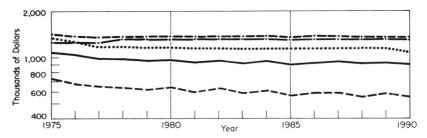

Figure 7-1a. Big Horn County educational expenditures and resources: Scenario I

········· Expenditures for 300 students per 1,000 population.
—·——· Receipts from gross proceeds, TIP & R&PP taxes, 40 mills.
——— Expenditures for 250 students per 1,000 population.
————— Receipts from real and personal property tax, 120 mills.
—·——· Receipts from gross proceeds, TIP & R&PP tax (lower bound)

the event that local revenues from the gross proceeds, industrial, and real and personal property taxes, along with the portion of the severance tax earmarked for local impact mitigation, were insufficient to support the incremental foundation program school expenditures associated with scenario I. It is clear, however, that the combination of revenue sources available to the county from property (mine, industrial, residential, and personal) is more than adequate at the county level, even with the minimum levy of 40 mills per dollar of taxable property. Expenditures indicated are at no time in excess of a million dollars for an estimated 250 students per thousand population ($1.1 million for 300 students per thousand population), whereas tax receipts are in the neighborhood of $1.25 million to $1.5 million, depending on whether the lower bound or medium estimate is used.

Were we to take as a minimum taxable property value the low estimates of taxable industrial property, low average real and personal taxable property value, and recalling that the price of coal for scenario I was given rather than estimated, thus leaving the gross proceeds tax yields unchanged, we would even so obtain revenues substantially in excess of incremental school expenditures. It is only for the case in which we assume all of the increase in enrollment would occur in town, while all but the taxable residential real and personal property would be located outside the urban school district, that we would find receipts fall short of expenditures even at a high of 120 mills per dollar of taxable property. But since the receipts from the mandated minimum levy against all taxable property in the county will exceed substantially the countywide educational requirements, and *if* this excess were available without restrictions for school equalization within the county, there would not be any problem with financing educational requirements at the scenario I level of coal development for Big Horn County. This obtains even under the most adverse set of assumptions. The situation does not change, for example, when we assume 300 students per thousand population, nor when this level of enrollment is compared with an assumed low yield from property taxes. One can say that largely because of the yield from the gross proceeds tax on the level of coal production in Big Horn County, proceeds from property taxation, even at a level of 40 mills per dollar of taxable property, would be adequate to finance education associated with the expansion of coal mining.

While the projected yields from the taxes on the increased property subject to taxation at the 40-mill mandated minimum levy alone would suffice to finance fully the increment in educational costs, this fact is not *directly* relevant to the matter of financing increased public services associated with coal development. The reason, of course, is that the yields from the 40-mill mandated minimum levy go to the county equalization fund for uses that are restricted to meeting the cost of the foundation programs of the poorer school districts within the county. Any monies in excess of the minimum needed to make up the differences between given school districts' requirements for the foundation level program and their tax receipts from the mandated 40-mill levy on property within their districts are destined for the state's equalization fund to underwrite the foundation program in other counties. The fact that yields from the 40-mill levy could meet all of the increased costs for education in the coal development counties is relevant only in the sense that it is likely to influence the Coal Board's response to the requests by affected communities for assistance from the local impact account provided for by the 1975 coal tax legislation. Thus, while the projected relationship between increased tax receipts and school expenditures does not address the issue directly, it is not irrelevant. We therefore review this relationship for each of the scenarios and for each county before evaluating the results in the contexts of potential transfers among different tax sources and accounts for financing educational services of affected communities.

Scenario I: Rosebud

In figure 7-1b are shown the receipt-school expenditure data for Rosebud County. Here, because of construction of Colstrip units 1 and 2, expenditures associated with a school enrollment that would reflect all construc-

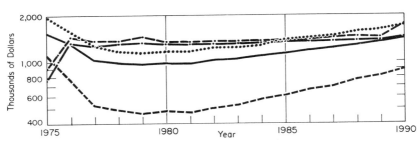

Figure 7-1b.   Rosebud County educational expenditures and resources: Scenario I

········· Expenditures for 300 students per 1,000 population.
—·——— · Receipts from gross proceeds, TIP & R&PP taxes, 40 mills.
————— Expenditures for 250 students per 1,000 population.
————— Receipts from real and personal property tax, 120 mills.
—·——— · Receipts from gross proceeds, TIP & R&PP tax (lower bound)

tion labor housed in the county are shown to exceed financial resources for the first year. After 1975, however, the resources for financing the estimated school enrollment would suffice to meet school expenditures even for the twin extreme assumptions of (1) 300 students per thousand population, and (2) the low estimates of taxable property coupled with the minimum levy of 40 mills. Again, if we have to assume that half of the student enrollment occurred in school district 4, which includes the town of Forsyth, and only half of the real and personal property and none of the mine and industrial property were taxable in that district, we would note a relationship similar to the educational expenditure lines and the receipts which correspond to a maximum of 120 mills per dollar of taxable property.[6] At no time during the ten-year period under review (1975–85) would school district revenues obtained from taxing residential real and personal property be equal to the implied educational expenditure. We recall, however, that a 40-mill levy is a mandatory minimum under law, and that any surplus obtained from this applied to mine and industrial property anywhere in the county would be available to support schools at least to the foundation level.[7] Moreover, tax revenues in the county, overall as in the case of Big Horn County, will exceed school expenditures after the first year. This is true even when the extreme

—————

[6] While both the expenditures and receipts would be reduced by half, the relationship between the two would be similar, that is, expenditures would exceed resources for educational purposes if they were dependent only upon tax revenues and educational obligations arising in school district 4.

[7] The quantitative implications of this are treated subsequently.

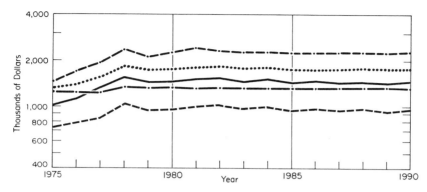

Figure 7-2a. Big Horn County educational expenditures and resources: Scenario II

········· Expenditures for 300 students per 1,000 population.
—·——· Receipts from gross proceeds, TIP & R&PP taxes, 40 mills.
————— Expenditures for 250 students per 1,000 population.
— — — — Receipts from real and personal property tax, 120 mills.
—·——· Receipts from gross proceeds, TIP & R&PP tax (lower bound)

assumption with respect to enrollment is coupled with the low estimate of tax yields. The combination of gross proceeds and industrial property taxes at the mandatory 40 mills will generate funds in the aggregate that exceed educational expenditures for the county at large.

Scenario II: Big Horn

In figure 7-2a we present information relevant to the set of conditions involving expansion of coal to 42 million tons in eastern Montana by 1980. The tax yield generated by the gross proceeds tax with a 40-mill levy would by itself provide sufficient revenue within the county to finance educational expenditures under the most extreme assumptions as to enrollment, valuation, and levy. It might be noted that while the residential real and personal property base could be much reduced in the event a large proportion of the population (and enrollment) was residing in Sheridan, Wyoming, none of the mine and industrial property base would be affected by this pattern of residential location. Basically, the gross proceeds tax generates sufficient revenue at the local level to leave unchanged the conclusion that revenues would be more than equal to sums required for educational purposes in Big Horn County irrespective of whether the population and enrollment associated with the mine employment is domiciled in the county or in Sheridan, Wyoming.

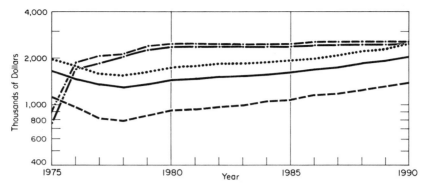

Figure 7-2b.   Rosebud County educational expenditures and resources: Scenario II

· · · · · · · · · Expenditures for 300 students per 1,000 population.
— · — — · Receipts from gross proceeds, TIP & R&PP taxes, 40 mills.
———————— Expenditures for 250 students per 1,000 population.
— — — — Receipts from real and personal property tax, 120 mills.
— · ——— · Receipts from gross proceeds, TIP & R&PP tax (lower bound)

Scenario II: Rosebud

School expenditures do not increase initially in scenario II over scenario I for Rosebud County (figure 7-2b) because the difference in construction employment due to mine expansion is dwarfed by employment involving power plant construction (Colstrip units 1 and 2) common to both scenarios. But the decline in employment after 1975 due to completion of the power plant is offset by expansion of employment in coal mine preparation and operation. As in the case of scenario I for Rosebud and both scenarios for Big Horn County, the revenues generated by taxes involving coal mining and related activities seem at least equal to the increment in educational expenditures induced by these activities. Extreme assumptions regarding the enrollment load, costs of financing the investment, the method of valuation adopted, and the level of the tax rate applied do not change the conclusion that property tax yields on mines and related facilities would by themselves generate enough revenues to finance education fully at the county level. Given the surplus overall county resources from a mandated minimum levy applied to mines and industrial properties, it is clear that financing of education would not be in difficulty even without state school equalization or the use of severance tax funds earmarked for local use, provided only that all such monies could be used to finance school programs, including those beyond the foundation level within the county. The data also indicate, however, that

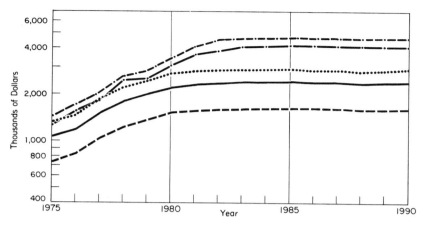

Figure 7-3a. Big Horn County educational expenditures and resources: Scenario III

· · · · · · · · Expenditures for 300 students per 1,000 population.
— · — — · Receipts from gross proceeds, TIP & R&PP taxes, 40 mills.
——————— Expenditures for 250 students per 1,000 population.
— — — — Receipts from real and personal property tax, 120 mills.
— · ——— · Receipts from gross proceeds, TIP & R&PP tax (lower bound)

were the school expenditures to be financed out of real and personal property taxes from school district properties alone, the affected school districts, such as 4 in which Forsyth is located, would have grave problems. Even at a maximum 120-mill levy, we note a deficiency of a half to three-quarters of a million dollars annually occurring in 1990.

Scenario III

Basically scenarios I–III represent "coal for export only" policies, and while there is some conversion of coal to electricity in every scenario due to the two 350 MW units at Colstrip, these have been justified as required for domestic (within-state) consumption. In figures 7-3a and 7-3b we find scenario III basic patterns and relationships among the tax receipt and school expenditure series that are quite similar to scenario II. Only the magnitudes involved differ substantially. This is understandable when we recall that we were postulating an increase in output from 20 million tons in 1975 for scenario I to 42 million tons in the case of scenario II. For scenario III, we have postulated six additional mines each of 10 million tons annual capacity for a total increment of 60 million tons, thus trebling the size of the increment.

The revenue yield from a 40-mill levy on Big Horn County taxable

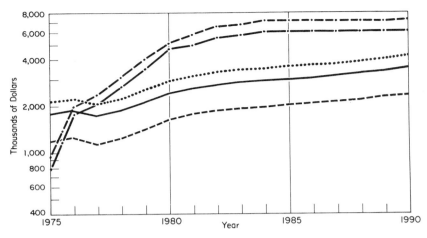

Figure 7-3b. Rosebud County educational expenditures and resources: Scenario III

········· Expenditures for 300 students per 1,000 population.
—·—— · Receipts from gross proceeds, TIP & R&PP taxes, 40 mills.
————— Expenditures for 250 students per 1,000 population.
— — — — Receipts from real and personal property tax, 120 mills.
—·——— · Receipts from gross proceeds, TIP & R&PP tax (lower bound)

mine, industrial, and residential real and personal property alone will exceed the estimated educational expenditures by an amount beginning with roughly a quarter of a million dollars in the initial year, a million by 1980, and by over two million by 1983 and thereafter. Not a great deal of employment, population, and hence school enrollment is associated with the mining of coal alone, while the property tax given by the gross proceeds and taxable industrial property formulas yields very large revenues.

While the initial year's educational expenditures in Rosebud County exceed revenue yields from a minimum 40-mill levy on only the taxable value of mines, industrial, and residential real and personal property,[8] the situation changes by the third year, yielding roughly a two million dollar surplus by 1980 and increasing additionally thereafter through 1985, at which time receipts exceed expenditures by around four million annually.

If we consider then the basic "coal for export only" scenarios, given

---

[8] Although the indicated excess of educational costs over revenue obtained from a minimum 40-mill levy on taxable property amounts to about a million, in scenario III for the first year, the amount of the severance tax yield earmarked for local impact even in the first year would exceed the excess in incremental educational expenditures associated with power plant construction at Colstrip.

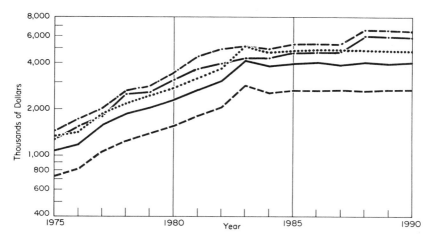

Figure 7-4a. Big Horn County educational expenditures and resources: Scenario IV

········· Expenditures for 300 students per 1,000 population.
—·——· Receipts from gross proceeds, TIP & R&PP taxes, 40 mills.
——————— Expenditures for 250 students per 1,000 population.
— — — — Receipts from real and personal property tax, 120 mills.
—·———· Receipts from gross proceeds, TIP & R&PP tax (lower bound)

the structure of Montana coal-related taxes, it is clear that revenues generated by the 40-mill minimum mandated are equal to implied expenditures on educational facilities and services.

Scenario IV: Big Horn

In our scenarios intending to illustrate a strategy that involved coal mining for energy conversion prior to export in Big Horn County, we have added to the two 10 million ton per year capacity mines of scenario III a 250-million cubic feet per day coal gasification plant. Since the gasification plant is not projected to be in operation before 1985, no additional industrial property becomes taxable before 1985 and only the estimated increase in real and personal property of the construction labor force beginning in 1981 will be reflected in the estimated incremental resources for educational purposes (figure 7-4a). Even with the facility coming into operation in 1985, it will be entered on the tax rolls at only 7 percent of its assessed value because it qualifies as a new industry for the first three years. It will not be until 1988 that there will be a significant increase in tax yields for this conversion scenario, compared with scenario III. In spite of this and the implied growth in school enrollment, receipts

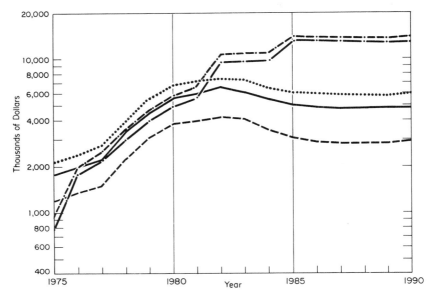

Figure 7-4b.   Rosebud County educational expenditures and resources: Scenario IV

· · · · · · · · · Expenditures for 300 students per 1,000 population.
— · — — · Receipts from gross proceeds, TIP & R&PP taxes, 40 mills.
————— Expenditures for 250 students per 1,000 population.
— — — — — Receipts from real and personal property tax, 120 mills.
— · ——— · Receipts from gross proceeds, TIP & R&PP tax (lower bound)

from the mandated 40-mill levy, countywide, exceed the aggregate incremental educational expenditures by a comfortable margin for all years with the possible exception of 1983, the year of peak enrollment. They begin with a quarter to over a half million dollar excess in the early years, increasing to over a million dollars between 1980 and 1987, and to over two million dollars thereafter.

Scenarios IV and IVa: Rosebud

In order to play out further the "energy conversion" strategy, we assumed that in addition to the coal gasification plant in Big Horn County there will be two 2,600 MW power plants in Rosebud County along with the four 10 million tons per year capacity mines appearing in scenario III. Moreover, in order to investigate what difference timing of power plant construction might make, we postulated two different schedules: (1) the first power plant begun in 1976 and coming on line in 1982, the second plant begun in 1979 and coming on line in 1985 (figure 7-4b), and

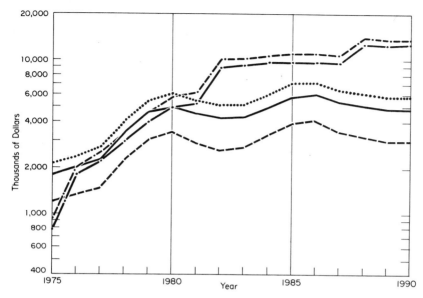

Figure 7-5. Rosebud County educational expenditures and resources: Scenario IVa

········· Expenditures for 300 students per 1,000 population.
— · — — · Receipts from gross proceeds, TIP & R&PP taxes, 40 mills.
————— Expenditures for 250 students per 1,000 population.
— — — — Receipts from real and personal property tax, 120 mills.
— · ——— · Receipts from gross proceeds, TIP & R&PP tax (lower bound)

(2) the first plant scheduled as before with the construction of the second delayed until 1982, coming on line in 1988 (figure 7-5). It is evident in both cases that growth in school enrollment associated with the large-scale construction activity in progress between 1975 through 1980 or '81, depending on the variant chosen, would involve annualized expenditures about equal to property tax receipts at the 40-mill levy. It is only after the construction activity has phased out that the receipts at 40 mills exceed the implied expenditures. When all construction activity has been phased out, and all mines and industrial plants are paying their full tax rate, the excess of receipts at the 40-mill levy will exceed expenditures by $8–10 million annually.

It should be noted in passing that the estimated upper bound enrollment figure, and hence expenditures, involves 300 students per thousand population among *construction workers*. Since the construction labor force represents one of constantly changing composition of skilled workmen as construction proceeds through various stages, these employees are more likely to be impermanent, and thus less likely to establish households at the site of construction.[9] The upper bound figures are likely to overstate school enrollment and expenditures. Expenditures corresponding to medium estimates of enrollment (250 students per thousand population) are in no case, following the first year, more than the minimum estimate of tax yields, and less than half as much after 1981.

ISSUES IN FINANCING EDUCATIONAL SERVICES

It follows that to the extent mine and industrial properties will not be distributed over the counties' school districts in proportion to the districts' enrollments, large differences between receipts from a 40-mill levy on properties within districts and the expenditures undertaken on behalf of foundation program educational services are almost certain to result. The fact that surplus receipts by school districts can be used for deficits in other school districts within the county to provide for the foundation program is of limited comfort. Indeed, since the foundation program is assured in any event by transfers from the state equalization fund, the large mine and industrial properties' tax payments benefit only the school districts in which such properties are located, and only in connection with financing of enriched programs beyond the foundation level.[10] The beneficiary of the expanded mine and industrial property in the eastern Montana coal counties would be, principally, the state's school equalization fund and the educational purposes that it serves throughout the state of Montana.

How large is the difference between the foundation program expenditure and total general fund budget? This of course varies from school district to school district for at least two reasons. First, the required expenditures per student to meet the foundation program will vary with circumstances, the size of school being one. Second, preferences among members of different school districts with reference to quality and cost of education will differ. Typically, the costs per student rise as the size of school decreases. The formula for determining the "foundation program

---

[9] We owe this point to Jack Schanz.

[10] An enriched program can be financed with a much smaller permissive or voted levy if the tax base is very large. Thus the residents of a particular school district with a large mine or industrial property will benefit in financing educational services. The statement in the text is only partly true, however, for another reason. Receipts from mine properties in the form of severance taxes provide for a fund to mitigate local impacts from coal development, including assistance with the cost of school facilities.

budget" for any given school involves a maximum lump sum for a given number of students, plus an increment, smaller than the initial average amount per student above the given number. For example,[11] for elementary schools with enrollments of 18 students or less, a lump sum of $8,149.50 represents a maximum lump sum amount with an incremental $314.50 per student in excess of 9 students. The formula works out to provide generally for a decreasing average amount per student as the size of school increases. For elementary schools exceeding 40 students but not more than 100, the average per student sum will range from $680.07 for a school with 41 students, down to $655.13 per student with an enrollment of 100, and $667.82 for the median number. For elementary schools between 101 and 300 students, the mandated expenditures will be $653.99 for a school of 101 students, $603.76 for 201, and $553.01 for 300. At the high school level the mandated amounts will figure roughly between $800 and $850 per student for schools in the 200 to 300 enrollment range, and $750 to $800 per student in the 300 to 600 enrollment range, with the lower figure representing the maximum for students in schools exceeding 600 enrollments.

As to the level of the total general budget, this will vary also from district to district, depending on the size of the combination of permissive and voted levies. An average for the state (1974–75) suggests a per student foundation budget of $613, with the permissive levy going to the maximum, that is, up to 25 percent of the foundation program, and voted levies as much as 38 percent of the foundation program expenditures. Of the total general fund budget, the foundation program in 1974–75 (with $613 per student) accounted for 61 percent, the permissive increment 15 percent, and the voted increase 23 percent, all rounded to the nearest unit.[12]

If the eastern Montana counties were to resemble the state average in this respect, roughly $400 per student per year would be the difference between the amount that was defrayed by the foundation program 40-mill levy and the total general fund budget per student. Capital outlays would be in addition to these amounts. Indeed, because of the incremental cost of new plant and equipment—costs that run much higher than the average for the facilities currently in place—and perhaps stiffer financing terms, the debt service charges on plant and equipment are

[11] The following data are taken from the formulas governing the 1974–75 school year.
[12] 1975 Montana Schools *Statistics,* Office of the Superintendent of Public Instruction, Helena, Montana, p. 7.

likely to run two to three times as much as the statewide average per student. Abstracting from the capital outlays that would be necessary, but including the annualized investment costs or debt service charges, the initial $400 per student would need to be about doubled.[13] Does it appear reasonable to expect this amount to be raised by additional permissive and voted levies in school districts which might have to bear the brunt of the increased service costs resulting from coal development, without simultaneously benefiting from mine and/or industrial property being added to their tax rolls? What individual school boards and district members might do is beyond our capacity to predict. However, if we review the data presented in figures 7-1 through 7-5, we observe that even at rates of *120 mills*—three times the 40-mill mandated levy—applied to real and personal property that does not include large mine and/or industrial properties, the financing of educational services would be in difficulty, with the difficulty greater the more intensive the development (i.e., with energy conversion as well as extraction in the activity mix).[14] This is likely to be the plight of towns such as Hardin, Forsyth, and possibly Hysham, in Treasure County, which may take some of the spillover population from the neighboring coal counties.

Now, it appears reasonable to assume that the financing of educational facilities and programs in districts which are affected by coal development not located within their taxing jurisdiction need not rely exclusively on the relatively meager tax base of residential and small commercial properties within their boundaries. The mine and industrial properties locating within the eastern Montana coal counties would, at the mandated 40-mill levy, yield tax revenues substantially in excess of implied educational expenditures in the coal counties. For example, the 40-mill levy on the mine and related industrial facilities, through the gross proceeds and industrial property taxes, supplemented by the modest additional property values, would provide the state equalization fund $2–3, depending on the development scenario considered, to support the statewide foundation program for every dollar that would be used in support of the foundation program within the two eastern Montana counties from which the funds originate. However, it must be noted that monies from

---

[13] See tables 7-1a and 7-1b for basic data.

[14] Actually, the first 40 mills of any district will yield the equivalent of the foundation program, for example, about $650 on average because any deficiency in the yield could be made up by one or another of the school equalization funds. Even so, the tax base in such cases would be insufficient even at 120-mill levies to provide for the accepted statewide standard of education.

the state equalization fund need not go to the counties of origin. This is not to suggest that the 40-mill mandated minimum levy and the educational equalization mechanism is inequitable in practice. Most would agree that it serves a worthy statewide purpose. But it could lead to problems for towns in the coal area that do not benefit directly from mine or related industrial properties unless it is used to transfer funds to them.

But in addition to the equalization fund, there is another source of revenue potentially available to make up the educational deficit—severance tax yields earmarked for local impact mitigation. A brief review of the magnitude involved suggests that these would be sufficient for this purpose, as well as for meeting the demands for local impact mitigation, under any one of the conditions postulated through our four scenarios.

However, it is necessary to recognize that for a variety of reasons the funding at issue is likely to be more a matter of the initial capital outlays than operating costs, and while capital might be raised through borrowing, with severance tax receipts shoring up the confidence of investors in the credit worthiness of the state, it is probably worth considering whether the portion of the receipts earmarked for local impact itself might not directly suffice for the investment in social overhead required by the conversion of the predominantly rural, stockman communities into a mining-industrial community structure. Accordingly, in tables 7-3a and 7-3b we show the cumulative annual investment in new educational facilities as a percentage of the cumulative accrual of severance tax yields earmarked for the local impact account computed at the maximum 17.5 percent rate through 1979 and 15 percent thereafter.

In tables 7-3a and b we display the array of possible outcomes, taking into account the potential variability in number of students per thousand population, giving thus a variation in the investment in educational facilities, on the one hand, and the variation in potential severance tax receipts based on our assumption of departure in the actual prices for coal by plus or minus 25 percent from our estimated price of $5 as the contract sales price for revenue estimating purposes. From discussions with knowledgeable persons in this area, it would seem that the number of students per thousand population is likely to be somewhat nearer our 200 estimate. The matter of the contract sales (relative) price expressed in terms of 1974 may be more or less than the $5 taken as the estimating value, but is likely to be contained within the range of plus or minus 25 percent. In any event, we have in table 7-3a and b the information necessary for checking the sensitivity of our results to the assumptions made with reference to the student enrollment and severance tax yields.

Table 7-3a. *Cumulative Investment in School Facilities as a Percentage of Cumulative Severance Tax Yields Earmarked for Local Impact Mitigation (Big Horn County)*

| | Students per thousand population | | | | | | | | | | | |
| | Scenario II | | | Scenario III | | | Scenario IV | | | Scenario IVa | | |
Year	200	250	300	200	250	300	200	250	300	200	250	300
Low yield estimates												
1975				0.88	1.10	1.33	0.88	1.10	1.33	0.88	1.10	1.33
1976				0.44	0.55	0.65	0.44	0.55	0.65	0.44	0.55	0.65
1977				0.36	0.45	0.54	0.36	0.45	0.54	0.36	0.45	0.54
1978				0.27	0.34	0.41	0.27	0.34	0.41	0.27	0.34	0.41
1979				0.22	0.28	0.33	0.22	0.28	0.33	0.22	0.28	0.33
1980				0.20	0.25	0.30	0.20	0.25	0.30	0.20	0.25	0.30
1985				0.09	0.12	0.14	0.16	0.20	0.24	0.16	0.20	0.24
1990				0.06	0.07	0.09	0.10	0.12	0.15	0.10	0.12	0.15
Medium yield estimates												
1975	0.88	1.10	1.33	0.88	1.10	1.33	0.88	1.10	1.33	0.88	1.10	1.33
1976	0.42	0.53	0.63	0.44	0.55	0.65	0.44	0.55	0.65	0.44	0.55	0.65
1977	0.29	0.36	0.44	0.34	0.43	0.51	0.34	0.43	0.51	0.34	0.43	0.51
1978	0.23	0.29	0.35	0.27	0.34	0.41	0.27	0.34	0.41	0.27	0.34	0.41
1979	0.18	0.23	0.27	0.22	0.27	0.33	0.22	0.27	0.33	0.22	0.27	0.33
1980	0.15	0.19	0.23	0.19	0.24	0.29	0.19	0.24	0.29	0.19	0.24	0.29
1985	0.08	0.10	0.12	0.08	0.11	0.13	0.15	0.18	0.22	0.15	0.18	0.22
1990	0.06	0.07	0.08	0.05	0.07	0.08	0.09	0.11	0.13	0.09	0.11	0.13
High yield estimates												
1975				0.88	1.10	1.33	0.88	1.10	1.33	0.88	1.10	1.33
1976				0.44	0.55	0.65	0.44	0.55	0.65	0.44	0.55	0.65
1977				0.34	0.43	0.51	0.34	0.43	0.51	0.34	0.43	0.51
1978				0.27	0.34	0.40	0.27	0.34	0.40	0.27	0.34	0.40
1979				0.21	0.27	0.32	0.21	0.27	0.32	0.21	0.27	0.32
1980				0.19	0.24	0.28	0.19	0.24	0.28	0.19	0.24	0.28
1985				0.08	0.10	0.12	0.13	0.17	0.20	0.13	0.17	0.20
1990				0.05	0.06	0.07	0.08	0.10	0.12	0.08	0.10	0.12

Table 7-3b. *Cumulative Investment in School Facilities as a Percentage of Cumulative Severance Tax Yields Earmarked for Local Impact Mitigation (Rosebud County)*

| | Students per thousand population | | | | | | | | | | | |
| | Scenario II | | | Scenario III | | | Scenario IV | | | Scenario IVa | | |
Year	200	250	300	200	250	300	200	250	300	200	250	300
Low yield estimates												
1975				3.48	4.34	5.21	3.48	4.34	5.21	3.48	4.34	5.21
1976				1.37	1.71	2.05	1.43	1.79	2.15	1.43	1.79	2.15
1977				0.73	0.91	1.09	0.86	1.07	1.29	0.86	1.07	1.29
1978				0.44	0.55	0.66	0.77	0.97	1.16	0.77	0.97	1.16
1979				0.32	0.40	0.47	0.69	0.86	1.03	0.68	0.85	1.02
1980				0.26	0.32	0.38	0.59	0.74	0.89	0.54	0.67	0.81
1985				0.11	0.13	0.16	0.22	0.28	0.33	0.21	0.26	0.32
1990				0.07	0.09	0.08	0.13	0.16	0.19	0.13	0.16	0.19
Medium yield estimates												
1975	3.28	4.11	4.93	3.48	4.34	5.21	3.48	4.34	5.21	3.48	4.34	5.21
1976	1.22	1.52	1.82	1.37	1.71	2.05	1.43	1.79	2.15	1.43	1.79	2.15
1977	0.68	0.85	1.02	0.72	0.90	1.07	0.84	1.05	1.26	0.84	1.05	1.26
1978	0.46	0.58	0.69	0.42	0.52	0.63	0.74	0.92	1.10	0.74	0.92	1.10
1979	0.34	0.42	0.50	0.29	0.36	0.44	0.63	0.79	0.95	0.62	0.78	0.94
1980	0.27	0.34	0.40	0.23	0.29	0.35	0.53	0.66	0.80	0.48	0.61	0.73
1985	0.13	0.17	0.20	0.09	0.11	0.14	0.19	0.23	0.28	0.18	0.22	0.27
1990	0.11	0.14	0.17	0.06	0.08	0.09	0.11	0.13	0.16	0.10	0.13	0.16
High yield estimates												
1975				3.48	4.34	5.21	3.48	4.34	5.21	3.48	4.34	5.21
1976				1.36	1.71	2.05	1.43	1.79	2.14	2.43	1.79	2.14
1977				0.70	0.88	1.06	0.83	1.03	1.24	0.83	1.03	1.24
1978				0.40	0.50	0.60	0.70	0.88	1.05	0.70	0.88	1.05
1979				0.27	0.34	0.41	0.59	0.74	0.88	0.58	0.72	0.87
1980				0.21	0.26	0.31	0.48	0.60	0.72	0.44	0.55	0.66
1985				0.08	0.10	0.12	0.16	0.20	0.24	0.15	0.19	0.23
1990				0.05	0.06	0.08	0.09	0.12	0.14	0.09	0.11	0.13

As we would expect, the "front end" capital costs of school facilities generally exceed the severance tax accruals for the first year in both Big Horn and Rosebud counties. Thereafter the accruals to the local impact fund from coal mined in Big Horn County exceed cumulative investment in school facilities, so that by 1980 investment will represent only from a fifth to less than a third of severance tax accruals, over the potential range of variation, and by 1990 only from 5 to less than 15 percent in the extreme.

For the conditions postulated in the several scenarios for Rosebud County, however, the capital stringency continues somewhat longer. In all cases, even under the most favorable set of circumstances, initial outlays for educational facilities required to service the construction-related population and school enrollment will exceed the severance tax accruals to the local impact account in the base year (1975) by a factor of 3 to 5. This, of course, is related to the construction of the two steam electric power units at Colstrip, and continues, albeit abated, through 1976 on the completion of construction of the power plant. Thereafter the cumulative severance tax accruals exceed the cumulative investment in school facilities in all cases except for the high enrollment assumptions for all scenarios in 1977. This condition persists into 1978 where the high enrollment assumptions for scenarios IV and IVa still produce "front end" capital outlays exceeding local impact account severance tax accruals. But again, as in the case of Big Horn County, by 1980 and thereafter, the accruals to the local impact account exceed capital outlays, and by a very substantial amount [ten to twenty times for the export scenario (III) and seven to eleven times for the conversion scenarios] by 1990.

Two conclusions may be drawn from this assessment. The first is that the severance tax yields are very ample, and the portion earmarked for the local impact account mounts rapidly as coal extraction proceeds. The second conclusion is that the front end costs begin partly in anticipation of the activities which produce the severance tax revenues. These capital outlays may be troublesome unless advances can be made by the state to local jurisdictions for the front end costs. It should be noted that while the funds earmarked for local impact are at most 17.5 percent of the severance tax yields, and as such would not be sufficient to provide adequate early year advances for local infrastructure investment, under the original legislation, 40 percent of the severance tax receipts were destined for general fund purposes. Whether or not these could be used is a matter we are not competent to address. One additional comment is warranted, however. The investment in school facilities is likely to represent an outside (upper bound) extreme. In reality, we do not expect the enrollment during the construction phase of energy development to be as large a proportion of the population as inferred from stable populations and, second, it is likely that a substantial part of the peak investment shown would be made unnecessary by a judicious use of mobile classroom facilities. With these considerations in mind, it appears that the revenues one could anticipate from the severance tax earmarked for local impact mitigation would be equal to coping with the problem.

SUMMARY AND CONCLUSIONS

In this chapter we set out to estimate what the implications of coal development in eastern Montana would be for the most significant single item in local government budgets—education. We first played out the implications of the increased population associated with increased employment and economic activity, for the school age population and school enrollment. This led to estimating the necessary investment in expanded facilities and the annual costs, both school operating costs and the estimate of appropriate annual debt service charges on the increment to school facilities.

Looking to the tax base to support this added educational program, we discovered that the mandated minimum 40-mill levy would provide revenues that would be generally in excess of any annualized costs associated with the increment in enrollment that the increment in coal development activities would occasion. However, the yields from the 40-mill levy are available within the county to support only the foundation program level education activities, which statewide average about 60 percent of the total cost of the general school budget. The remainder of the tax yield, varying between two and three times the amount that would be retained for the coal counties' foundation programs, would be destined for the state equalization fund to help support the foundation program statewide. Accordingly, although the yield from the school district levies of 40 mills would provide resources that would be adequate to cover all of the educational expenditures, were they permitted to, the excess costs over the amounts allowable under the foundation program would create serious difficulties for the school districts, that is, those located within the towns of Hardin, Forsyth, and possibly Hysham in Treasure County, that are likely to bear the brunt of the demand for educational services without having the large mine and related industrial properties on their tax rolls.

The difficulties created by the dissociation between taxable property and the location (jurisdiction) within which the demands for this public service would be made can be resolved by the use of state severance tax receipts earmarked for mitigating local impacts resulting from coal development. A comparison of cumulative investment requirements with cumulative earmarked tax receipts suggests that, particularly after the first year or so, the former would represent only a small fraction of the latter for the coal for export only scenarios, and a somewhat larger fraction of the latter for the energy conversion scenarios. Accordingly, the financing of the most significant area of community services does not appear to be a serious problem no matter what level of intensity up to the limits we have investigated—and these limits represent double or treble the level currently being implemented (scenario II). It is important to note, however, that this outcome is a result of the terms of the severance tax which provides for local impact mitigation funds. The tax, however, is related to the extraction, not the conversion, of coal; and while the conversion and the facilities used for this purpose are also taxed, the terms with respect to the disposition of the revenues differ, that is, they make no allowance for mitigating local impacts. At the moment, given the projected receipts of severance taxes on the mining of coal, and the likelihood that there would be an increment in coal extraction corresponding to any increment in coal conversion, the severance tax provisions may continue to suffice. This is obviously an area which should receive close attention by those responsible for administration and legislation in the area.

APPENDIX B

SCHOOL POPULATION MODEL

by John V. Krutilla and V. Kerry Smith

This model estimates increments to the school population in a given area for elementary and high school separately, derives the required investments, and computes the annualized costs. It is assumed that a fixed proportion of the total school population is in elementary grades and in high school.

Equation (1) defines the estimation procedure for the elementary school population

$$ESPOP_t = ESPOP_{t-1} + y^z \widehat{\Delta POP}_t \tag{1}$$

where $ESPOP_t$ = elementary school population in period t

y^z = proportion of the population that were elementary school students for scenario Z. This figure was estimated at 19 percent of the population

$\widehat{\Delta POP}_t$ = estimated increase in population in period t (derived from the Harris model)

Similarly, the high school population estimate is based on the population forecasts of the Harris model as given in equation (2)

$$HSPOP_t = HSPOP_{t-1} + \bar{y}^z \widehat{\Delta POP}_t \tag{2}$$

where $HSPOP_t$ = high school population in period t

$\bar{y}^z$ = proportion of the population that were high school students for scenario Z

These estimates are utilized to derive the investments necessary to accommodate the increases in the student population. These investments are based on the increment in each student population over the previous peaks, as defined by equations (3) and (4).

$$MESPOP_t = \text{Max}[ESPOP_1, \ldots, ESPOP_{t-1}] \tag{3}$$

$$MHSPOP_t = \text{Max}[HSPOP_1, \ldots, HSPOP_{t-1}] \tag{4}$$

The required investments are derived as an average requirement per pupil for elementary and high school students as given in equation (5)

$$I_t = a_1[ESPOP_t - MESPOP_t] + a_2[HSPOP_t - MHSPOP_t] \tag{5}$$

$$\text{for } ESPOP_t - MESPOP_t > 0$$

$$HSPOP_t - MHSPOP_t > 0$$

where a_1 = dollar investment required per elementary school student

a_2 = dollar investment required per high school student

The annualized capital cost associated with the cumulative investments necessary in period t (ACC_t) with a loan term, T, and interest rate, r, is then given by (6)

$$ACC_t = \frac{r}{(1 + r)^T - 1} \left(\sum_{i=1}^{t} I_i \right) \tag{6}$$

8 Community Services and Expenditures—Noneducational

INTRODUCTION

The bulk of the expenditures for tax-financed public services provided by local governments in Montana go for education. Indeed, when expenditures are segregated in order to separate capital from current accounts,[1] and self-liquidating utility services are removed so that only publicly financed (nonpriced) services are included, something like 60 percent of the local government expenditures, taking the average over all local taxing jurisdictions in Montana, are accounted for by educational services. For this reason we treated educational services at the local level separately in chapter 7.

When only current accounts are considered, again, county government (noneducational) services will account for roughly only about 23 percent of total local government costs in spite of the fact that there are some sixty-odd service and administrative expenditure categories listed in the State Commission on Local Governments' accounts.[2] The proportion of the total accounted for by towns and cities represents the remaining 17 percent.

---

[1] Capital outlays in a given year may distort the comparisons and are thus removed for purposes of a given year's comparison, although debt service charges, corresponding to annual or current account equivalents, are included.

[2] Unpublished data supplied by courtesy of Steve Turkiewicz of the State Commission on Local Government, Helena, Montana.

Because of the large number of services other than educational that individually represent so small a proportion of total local government expenditure, we will not undertake an analysis of each service category as we did for education. Not only would such detailed analysis be beyond the scope of this study, it is not entirely clear that we could justify the analytical effort for our purpose. Given the same tax base available to finance county services as is available to finance educational services, unless there is some reason to think other services tend to increase disproportionately with development, we can feel some assurance regarding the adequacy of tax revenues to finance county functions. Moreover, those services—often requiring new capital facilities such as enlarged, improved and extended roads—are frequently aided by intergovernmental transfers handled outside county tax and public service institutions. For our purpose, at least, a coarser grained, more aggregative analysis is expected to suffice.

In this chapter then, we go beyond our preliminary analyses of chapter 6 for a more detailed attempt to estimate the changes in the size of tax-financed public expenditures that would attend coal extraction and related activities for each of the affected counties for the different coal development strategies. *We do this by undertaking a cross-sectional analysis of the local government expenditures that have been provided by the Census of Governments (1972).*[3] Having estimated the expected per capita tax-financed expenditures for county purposes, we provide estimates of per capita tax revenues, based on a set of variables found to influence the tax rates that would be relevant to our purpose. The analysis is next applied to the city or town expenditure and receipts per capita to distinguish the conditions that are expected to obtain at the town level from those that would be expected at the county level. We begin, then, with the analysis of per capita expenditures found among Montana counties.

ESTIMATING THE EXPENDITURES OF COUNTY BUDGETS

The growth in population that has occurred in the coal counties, particularly in Rosebud as result of the increased level of coal mining and the construction of the two new generating units at the Colstrip steam electric generating plant site, has already affected the demand for public

---

[3] U.S. Bureau of the Census, *1972 Census of Governments,* vol. 4, *Government Finances.*

services. This level of expansion, however, corresponds to the much more modest levels of scenarios I and II. With the sort of expansion that would attend either scenarios III or IV, one would anticipate significant changes in community size and structure, and these changes in turn are likely to have significant implications for per capita expenditures for public services at the county level. It is conceivable, for example, that at the higher levels of expansion the composition of the county population would change, with new migrants reflecting different preferences for public services and, indeed, preferences for higher levels and enlarged choice of public services. In some respects, quite independent of preferences, the changes that would be expected to occur as a result of changed density and diversity of population may require, for example, among other things, higher per capita expenditures for public safety to provide for the same level of protection. For a variety of reasons, then, there is expected to be an increased demand for public services compared with those sought from a more self-sufficient, sparsely populated, predominantly cattle-ranching community.

In order to reflect the differences in the per capita demand for tax-financed public services that may be related to the factors mentioned above, we can do a cross-sectional analysis of the 56 counties of Montana. Here we find counties of various sizes, and community diversity, ranging from some like Petroleum or Golden Valley counties with less than a thousand persons to Yellowstone County, having close to a hundred thousand. This range of size and diversity will bracket the range within which Big Horn and Rosebud counties could be expected to fall, irrespective of the particular scenario considered. Below, we present the results of regressing per capita expenditures for county government purposes (excluding educational expenditures, and capital outlays, but including debt service charges) on per capita income, per capita taxable property value, population density, and area.

$$X_1 = 38.071 + 0.01202\ X_2 - 0.32687\ X_3 + 0.02143\ X_4 - 0.0080\ X_5$$
$$\qquad\qquad (2.0958)\qquad (0.62109)\qquad (4.5459)\qquad (-2.0605)$$

$$R^2 = 0.478641 \qquad \text{S.E.} = 35.28666$$

where: X_1 = per capita county expenditure on noneducational tax-financed services in dollars

X_2 = per capita income in dollars

X_3 = population per square mile

X_4 = per capita taxable value of property in dollars

X_5 = area in square miles

The values in the parentheses below the estimated coefficients are Student t statistics for the null hypothesis of no association between each independent variable and the dependent variable.[4]

The regression equation indicates that expenditures tend to rise with per capita income ($12.02 per thousand dollars), are negatively related to population density (indeed, the highest expenditures per capita tend to be associated with the most sparsely populated counties), are positively related to the value of taxable property ($21.43 per thousand dollars), and are negatively related to area. It is not surprising that the demand for public services, hence expenditures, should rise with income and wealth (X_2 and X_4); nor that some economies of scale in providing services may be achieved within the size range of Montana counties that would result in reduced expenditures per capita with increased density and scale of service. The negative association with area is not intuitively obvious, although time permitting, it would probably have been instructive to have computed the area of each county net of federal lands.

Using the equation, we can obtain predicted value of expenditures per capita for future years for each of the several scenarios for the two counties. These are given in tables 8-1a for Big Horn County and 8-1b for Rosebud. It should be kept in mind that these expenditures are intended to reflect *tax-financed* public expenditures exclusive of education, and of course, capital outlays.

If we compare the expenditures per capita for Big Horn and Rosebud counties, we find that for scenarios I and II the difference in the estimated expenditures per capita is related primarily to the difference in per capita incomes between the two counties. Indeed, per capita income for scenario I for Rosebud exceeds projected per capita income for any of the scenarios

---

[4] Population density is retained in the expenditure equation despite its low t value for several reasons. There are economic reasons for suspecting that the demographic characteristics and spatial configuration will affect the cost of providing particular services. The sign of the estimated coefficient agrees with *a priori* expectations for a number of such services. Equally important, it is necessary to consider the consequences of omitting a variable from a regression equation in terms of the effects on the properties of the estimator on the remaining coefficients. That is, if we were to incorrectly exclude X_3 from the equation (i.e., the true coefficient is nonzero), then we run the risk of biasing our estimates of the coefficients for the remaining variables. By contrast, if we retain X_3 when its true effect on X_1 is zero, then we may reduce the precision with which the remaining effects may be measured. Accordingly, data-based decisions on the specification of econometric models involve a tradeoff between the likely bias and variance that will be accepted. For the present purpose, where bias may well present a problem, we have elected to retain variables in the estimated equations since there are reasonable theoretical arguments for their inclusion regardless of the results of conventional t tests.

Table 8-1a. Estimated Per Capita County Expenditures, Big Horn County, 1975–90
(dollars)

Year	Scenario I	Scenario II	Scenario III	Scenario IV	Scenario IVa
1975	139.07	139.07	139.07	139.07	139.07
1976	142.20	151.55	151.07	151.07	151.07
1977	143.29	158.33	156.26	156.27	156.27
1978	144.34	170.31	174.65	174.65	174.65
1979	145.88	164.39	177.94	177.96	177.96
1980	145.79	172.07	191.65	191.67	191.67
1981	147.64	171.38	213.44	208.77	208.76
1982	147.14	170.68	218.61	206.87	206.86
1983	148.91	172.85	230.17	202.54	202.53
1984	148.66	172.42	230.22	208.00	208.01
1985	150.59	174.74	229.89	216.08	216.11
1986	150.22	174.15	232.26	215.29	215.32
1987	150.42	176.02	231.20	217.48	217.49
1988	151.98	175.42	233.53	246.38	246.39
1989	151.61	177.25	232.41	248.53	248.56
1990	153.15	176.60	234.64	247.17	247.18

for Big Horn (see figures 3-4a and 3-4b). While the absolute difference between the per capita incomes for the two counties never declines, the relative per capita income difference narrows over time. However, with the widening difference in per capita taxable property value resulting from our postulated siting of mines and facilities (scenarios III and IV), the difference in the estimated per capita expenditures grows for the two counties. Had we postulated alternatively the siting of the power plants in Big Horn rather than Rosebud County, the per capita expenditure differences between the two would have been reduced, and perhaps even reversed.

In any event, it is important to appreciate that the *estimated* expenditure per capita depends to a very large extent on the variable for per capita taxable value of property in the estimating equation and that the per capita taxable value in our upper bound scenarios may well fall outside the range that is accurately predicted by the equation. With this possibility, it may be useful to review briefly the range of values in the 1972 Census of Governments for tax-financed county expenditures per capita. Of the 56 counties, there were four with expenditures per capita in the range of $200 to $250, excluding educational expenditures and capital outlays. (Big Horn and Rosebud counties had expenditures per capita of $54.85 and $105.62 respectively.) Another six had expenditures per

Table 8-1b. *Estimated Per Capita County Expenditures, Rosebud County,*
1975–90
(dollars)

Year	Scenario I	Scenario II	Scenario III	Scenario IV	Scenario IVa
1975	144.47	144.47	143.07	143.07	143.07
1976	190.72	204.53	204.53	192.87	191.16
1977	202.66	224.98	221.19	209.69	209.69
1978	206.69	233.89	252.92	218.30	218.30
1979	209.76	244.06	286.99	227.91	229.07
1980	209.74	245.72	318.21	240.47	247.11
1981	211.55	246.23	337.96	252.45	281.63
1982	210.50	244.26	359.69	339.25	416.66
1983	211.01	244.85	364.24	351.65	421.02
1984	209.13	242.64	373.88	387.19	394.62
1985	209.27	242.25	371.18	502.25	364.78
1986	207.23	239.56	370.16	518.36	362.02
1987	206.91	239.05	365.16	522.70	388.08
1988	204.68	236.26	362.93	523.37	493.34
1989	204.17	235.40	357.14	526.23	509.89
1990	201.68	232.34	354.16	520.00	511.02

capita of between $150 and $200. It is interesting to observe that almost
all of the above represent counties with very small populations, roughly
2,500 persons or less. It is likely that counties with such sparse populations have no urban centers and that the public services normally provided in urban settings with some economies of scale, and which appear
in the expenditures of towns and cities, are provided in part by these
county governments. All but one of the counties with populations over
15,000 by contrast had per capita expenditures in the $50 and $85 range,
with most nearer the lower end of the range. Since this is the range class
into which Big Horn and Rosebud would fall were coal development to
proceed in a way corresponding to scenarios III, IV, or IVa, the predicted per capita expenditures obtained from our estimating equation appear to be upper bound values, if not extreme.

Before we conclude in this fashion, several caveats appear to be in
order. Data referred to above represented information obtained from the
1972 Census of Governments, and we have reason to suppose that more
recent data on these counties would reflect a substantially higher average.
But even with double digit inflation between 1972 and 1974 in whose
constant value dollars our computations are carried out, more recent
data would hardly show expenditures per capita of three to four times
1972 levels.

An equally important consideration is the difference between the case where the bulk of one's public facilities has been obtained at preinflation costs, with only operating expenditures reflecting the inflation, and the case in which the bulk of the cost of *facilities* as well as operation and maintenance expenditures are to be incurred in the postinflation period. The costs of local governments whose populations and level of public services remain relatively stable over this period are likely to be substantially less per capita, in the short to intermediate run, than the costs of those jurisdictions which must obtain the bulk of their facilities after the sharp rise in costs. With this factor likely to be a very significant one, it is not possible to rely exclusively on the results of cross-sectional analysis of 1972 data. On the other hand, in the previous chapter we used 1974 construction costs to predict new facilities that will be required to service the educational needs of the additional population associated with the postulated coal extraction and conversion activities. Even with facilities designed to meet modern requirements of space for school children of different age classes, and conservative estimates of construction costs per square foot, capital investment and corresponding current costs (or debt service charges), and school operating and maintenance costs relevant to that period, we inferred that the mandated *minimum* levy would be equal to the total educational costs if it could be used entirely within the counties for such purposes.[5] In short, even at the level of current replacement, maintenance, and operating costs, the revenues that would be expected to be yielded from the greatly enlarged tax base would be adequate *at a fixed mill rate* for the new facilities and their operations.

The problem with relying on a fixed tax rate is that, except for the case of school districts, the tax rate, when fixed, typically has an upper bound or is a maximum rather than a minimum rate. Generally speaking, when a local government adds to its tax rolls a large new industrial property, there are few things more likely to be reduced, other things remaining equal, than the tax rate. On the other hand, such properties rarely bring in their wake the kind of large structural changes that can be anticipated with the more intensive development scenarios that we have postulated. Accordingly, it is necessary to examine this question at least briefly in order to obtain some idea of the revenues which could be obtained, taking into account the possibility of an inverse relation between the tax rates and the aggregate taxable value of property.

[5] Recall that while the mandated minimum levy would produce revenues sufficient to cover current costs, it would not be available to the counties in amounts in excess of the needs to meet the (minimum) foundation program.

ESTIMATING THE REVENUES OF
COUNTY BUDGETS

Since we are separating tax-supported public services from self-liquidating publicly provided utility services, a convenient breakdown of major functions on which data are usefully assembled will be found in the reports of the State Department of Revenue[6] or the publications of the Montana Taxpayers Association.[7] Here a dozen functional categories are treated, giving the tax rate for each, the general fund millage rate and the total rate, along with the taxable property by county.

In our treatment of the estimated tax revenue available for county and local purposes in chapter 6, we selected rather conservative estimates of levies to apply to taxable property, for reasons given there. For the school districts we assumed that only the legal minimum 40-mill levy would be used. Similarly, for the county levies we assumed 27, 31, and 36 mills. The average total levy for the eastern Montana coal counties in fiscal 1974 was 36 mills, with Big Horn County's levy 26.72 mills and Rosebud County's 34.38. The average for all counties in Montana was about 45 mills. These levies lie within the lowest quintile for Montana counties, and it may even be hypothesized that the levies could increase, or perhaps would be required to increase, in order to provide the public services indicated by the changes in demographic and community structures induced by coal development.

We recognize that we are likely to encounter opposing tendencies regarding the level at which the tax rate will be set under the conditions postulated. On the one hand, as large industrial properties become subject to local property taxation, the tendency in many communities is to lower the rate because the increased tax base will provide revenue for essential services while reducing property taxes for existing local residents. However, when we contemplate development in an essentially rural, sparsely populated area, which would result in substantial population growth and related changes, preferences for public services may change.

Recognizing that the preexisting rates, and even the range within which we elected to make our first-pass estimates may be subject to amendment, we need to get a better idea of how tax rates vary as a func-

---

[6] The *Report of the State Department of Revenue to the Governor and Members of the Forty-fourth Legislative Assembly of the State of Montana* for the period July 1, 1972 to June 30, 1975 is, at the time of this writing, the most recent available.

[7] Montana Taxpayers Association, *Montana Property Taxation, 1975,* is an equivalent publication treating property taxation during the same biennium.

tion of scale, or the size of population, and of preferences reflected in population density, income per capita, tax base, and similar factors. Accordingly, we postulate that the rate for each of the categories of public service[8] would be a function of population, population density, personal income per capita, the taxable value of property, and the area in each county. Since not all counties levied taxes for every one of the dozen functional categories (see footnote 8), only those counties showing an entry were selected as the sample. That is, the regression analysis reflected the fact that service was not being offered where no rate was shown explicitly in connection with a particular category of expenditures.

If one were to expect a large systematic component in the variation among counties in the rate levied as a function of the variables examined, the expectation would not be realized. As one might infer, there was colinearity between the population and population density variables. Population density was selected since it was more consistently statistically significant at the 0.95 level than population. Income, the per capita value of the tax base, and area were each statistically significant at the 0.95 level. However, only one independent variable was statistically significant in any one of the regression equations. The levies for supporting county agents were an exception since we found a negative significant association between population density, area, and the per capita taxable value of property. Where statistically significant variables emerge, then, the estimating equation for tax rates was, with only one exception, a function of only one variable, with that variable often differing according to functional category.

It is probable that if a more thorough, detailed analysis were performed, one might obtain better estimating equations for the millage levy in relation to some, if not all, of the special county government service categories. However, our purpose was to run a rough check on whether there is a strong relation between county tax rate by specific service category and the scale, community structure, and/or the tax base per capita. To the extent our variables picked up the scale and community structure components within each county government service category, it would appear that random elements affecting the tax rates tended to dominate any systematic component. With so little promise of estimating the total rate based on summing the estimated rates in the specific service categor-

---

[8] These were, in addition to general fund, the special budget categories of roads, bridges, poor, extension agent, county fair, library, airport, weed control, cemeteries, debt service, senior citizens, and "others." Data corresponded to fiscal year 1975.

ies, it appears sensible to work with just the total levy or tax rate to meet the total of the traditional county government services. What we do, then, is determine the coefficients of an equation estimating county tax rate by using a multiple regression having the rate expressed as a function of income per capita, population density, value of taxable property per capita, and area. Indeed, the population density and area variables were not statistically significant at the 0.90 level, but were significant in connection with some of the separate functional categories and were thus retained, based on the criterion of maximizing the explained variance. Despite its relatively low explanatory power (19 percent of the variation), the following equation is used in subsequent analysis:

$$X_1 = 43.242 + 0.001708 \ X_2 + 0.11824 \ X_3 - 0.001294 \ X_4 - 0.01407 \ X_5$$
$$ (2.6812) (0.9811) (-2.3407) (-1.4985)$$

$$R^2 = 0.18965, \quad \text{S.E.} = 8.67688$$

where: X_1 = tax rate in mills
 X_2 = per capita income in dollars
 X_3 = population per square mile
 X_4 = per capita value of taxable property in dollars
 X_5 = area in square miles

The values in the parentheses below the estimated coefficients are Student t statistics for the null hypothesis of no association between separate independent variables and the dependent variable. Although the coefficient of X_3 is not statistically significant, we elect to retain the variable to maintain parallel specification with our expenditure equation.

As previously mentioned, the tax rates in Big Horn and Rosebud counties in 1974 were respectively 27 and 34 mills. Using our estimating equation, one would expect something closer to 39 and 43 mills as the total levy for the two counties respectively, abstracting from the random elements in the determination of the tax rates.[9] If we take these as the bases for the estimated changes in total levies for county government services for each of the five scenarios, there will not be much variation in tax rates over time due to the combination of changes projected in population, tax base, and personal income per capita. Because all of the variables in the estimating equation will remain virtually constant in scenario I for Big Horn County except for income per capita (assumed to

---

[9] Recall that Big Horn and Rosebud counties were in the lowest quintile of counties ranked by tax rates.

increase because of projected productivity gains per worker) and the tax rate is positively related to income per capita, it is estimated to increase by about 3 percent by 1985 and 4 percent by 1990. In none of the other scenarios does the tax rate increase for Big Horn County, since the negative influence on tax rates of growing property value per capita offsets all of the other positive influences. In scenario II the rate remains virtually constant but the estimated rate in scenarios III–IVa falls by 4 percent by 1980 and ultimately some 8 percent by 1990, with a low point in the vicinity of 34 mills.

Because the development scenarios postulated for Rosebud (along with the actual expansion of steam electric power generation at Colstrip) represent a substantially greater increase in industrial property values, the variation from the base period is greater here. The estimated base year rate of 40.42 would decline in each of the development scenarios, falling ultimately to 32.66 mills in scenario III, 23.57 mills in scenario IV, and 24.22 mills in scenario IVa, as the industrial property value per capita increases. The influence of the inverse relation between tax rate and value of property per capita completely dominates the other factors (population density and income per capita) to which the tax rate is positively related.

Since the systematic component of variation given by our estimating equation is not particularly large, it is perhaps useful to address the random variation in the tax rate. Given the variance associated with our estimated relationship between independent and dependent variables, we would expect that 95 percent of the time the actual tax rate would fall within a range of plus or minus 6 mills of the estimated rate for scenarios I and II for Big Horn County, and between plus or minus 6–8 mills, depending on the year (larger the further out in time, or greater the estimated property value per capita) for scenarios III–IVa. Similarly, the interval size at the 95 percent confidence level for Rosebud County varies between plus or minus 5–7 mills, depending on the year for scenarios I and II, to between plus or minus 5–19 mills, depending on the year, for scenarios III–IVa. Despite the potential variation in tax rates in later years given by our estimating equation, it will nonetheless provide the best initial estimates for our purposes, based on Montana experience.

Using the estimated tax rates by year, we can apply them to the corresponding tax base. For this purpose, we take the medium estimates for the tax base as described in chapters 5 and 6. Briefly, this assumes established contract prices for coal for the scenario I and II outputs, and an

Table 8-2a. Estimated Per Capita County Revenues, Big Horn County, 1975–90 (dollars)

Year	Scenario I	Scenario II	Scenario III	Scenario IV	Scenario IVa
1975	153.92	153.92	153.92	153.92	153.92
1976	156.03	170.45	169.82	169.82	169.82
1977	156.96	180.20	177.42	177.42	177.42
1978	157.47	196.96	202.66	202.66	202.66
1979	158.54	187.67	207.45	207.46	207.46
1980	158.11	197.41	226.16	226.18	226.18
1981	159.53	196.44	253.65	248.08	248.08
1982	158.81	195.49	259.91	246.44	246.43
1983	160.14	197.47	273.14	240.64	240.63
1984	159.70	196.82	273.23	247.90	247.90
1985	161.19	198.96	275.85	257.82	257.84
1986	160.65	198.15	275.29	256.85	256.88
1987	160.61	199.83	274.14	259.13	259.14
1988	161.77	199.01	276.56	293.42	293.43
1989	161.22	200.63	275.34	295.73	295.73
1990	162.32	199.73	277.63	294.30	294.31

estimated $5 per ton for coal corresponding to our scenario III output, in computing the property value using the gross proceeds method. Taxable industrial property is figured at about 7.5 percent of investment (see pages 75 and 92), and the value of real and personal property is computed on the medium estimate of $2,000 per capita (see pages 96–98). Given the estimated tax rates and the medium tax base, we show in table 8-2a the estimated per capita revenue yields for Big Horn County.

For convenience we present the revenue data of table 8-2a, along with expenditure data from table 8-1a in figures 8-1 and 8-2.

Reviewing the data, it is apparent that the combination of estimated rates and medium tax base would provide revenues that would be equal to the projected expenditures and would place Big Horn County in the top quintile of Montana counties, and indeed among the very highest of comparable populations and densities, as would be anticipated for the higher development scenarios. This, of course, is not a surprising outcome, in light of the results of our analysis of tax yields in chapters 6 and 7.

While it seems that the county tax yields will cover current accounts, even the annualized costs of capital outlays (since debt service charges were included in per capita expenditures), this does not say that annual revenues will be sufficient to support capital outlays as they occur. Need will doubtless exist for borrowing, or some other form of capital advance.

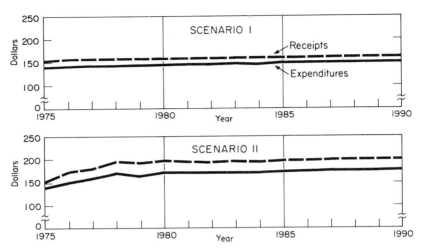

Figure 8-1. Estimated county per capita expenditures and receipts for Big Horn County, 1975–89: Scenarios I and II

What is indicated, however, is that the yields would be sufficient to meet all operating and maintenance costs, along with the interest and amortization on capital improvements for which borrowing was required.

Another observation might be worthy of note. While the more intensive the development, the greater the feasible level of expenditures (for any given tax rate), it does not follow that it is always greater for scenarios IV and IVa than it is for scenario III. The reason, of course, is that a very large proportion of the total property tax is related to the gross proceeds from mining, which are independent of whether the given volume of coal is converted locally or out of state. The increment in property taxes resulting from the investment in energy conversion facilities and the additional nonindustrial real and personal property tax base associated with scenarios IV and IVa is not proportional to the increase in population during the construction phase and exceeds the per capita yields of scenario III only modestly thereafter.

Proceeding now to Rosebud County, we give for the estimated medium tax base and our projected tax rates the estimated per capita tax yields in table 8-2b.

The estimated county tax yields shown in table 8-2b indicate that for scenarios I–III, tax yields exceed estimated expenditures per capita for all the years shown. In scenario IV, expenditures appear to exceed yields by a modest amount beginning in 1985 and for scenario IVa be-

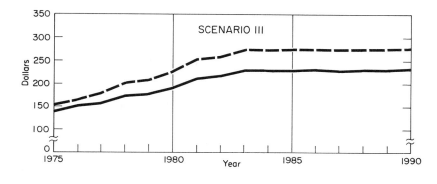

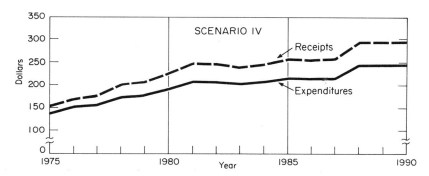

Figure 8-2. Estimated county per capita expenditures and receipts for Big Horn County, 1975–89: Scenarios III and IV

ginning in 1988 (compare tables 8-1b and 8-2b). This is better observed in figures 8-3 to 8-5.

These results must, however, be kept in perspective. Since the expenditure and receipt sides of the budgets are not made independently of each other, it is not likely that the county will be running large surpluses in the early years—or at any time for that matter. Similarly, if what was thought to be required by way of local expenditures were larger than the projected receipts, *at the projected tax rates,* the rates could be expected to be adjusted to provide the necessary receipts. Because of the very ample tax base compared with that of Montana counties generally, the projected rates in the period 1985–90 are roughly only half the level of tax rates which are traditionally accepted within the local tax environment among Montana counties. Indeed, if the predevelopment tax rates (42 mills in Big Horn County in fiscal year 1973, and 46 mills in Rosebud) were used, the per capita tax yields in Big Horn County would ap-

Table 8-2b. Estimated Per Capita County Revenues, Rosebud County, 1975–90
(dollars)

Year	Scenario I	Scenario II	Scenario III	Scenario IV	Scenario IVa
1975	149.92	149.92	148.24	148.24	148.24
1976	222.13	242.66	227.00	224.37	224.37
1977	238.86	270.47	265.51	248.12	248.12
1978	243.30	281.27	305.79	258.26	258.26
1979	246.26	293.72	345.62	271.63	273.27
1980	245.49	295.58	377.91	289.28	298.50
1981	246.92	295.74	396.05	305.95	341.08
1982	244.92	292.97	413.74	402.31	446.55
1983	244.41	293.10	417.50	413.46	453.97
1984	241.16	289.89	425.63	437.09	445.61
1985	240.08	288.61	424.42	473.29	427.69
1986	236.58	284.69	424.97	476.02	425.54
1987	234.87	283.26	421.81	479.72	442.49
1988	231.02	279.11	421.34	482.35	480.00
1989	228.93	277.11	417.20	486.41	485.99
1990	224.67	272.47	415.79	488.25	489.63

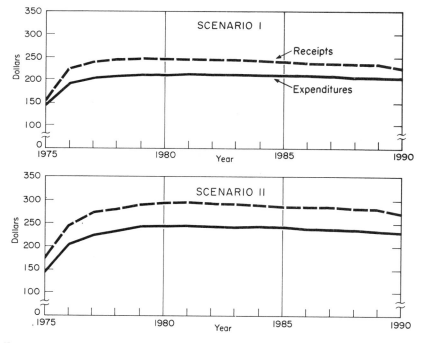

Figure 8-3. Estimated county per capita expenditures and receipts for Rosebud County, 1975–89: Scenarios I and II

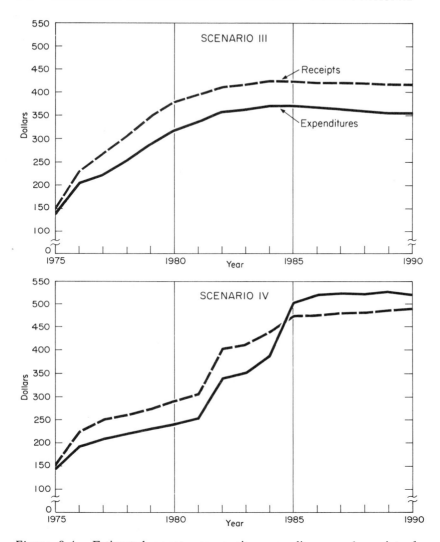

Figure 8-4. Estimated county per capita expenditures and receipts for Rosebud County, 1975–89: Scenarios III and IV

proach $400 and in Rosebud County would exceed $900 in the later years for scenario IV. In short, given the tax bases involved in the more intensive development scenarios, the yield from tax rates at levels up to what is traditionally accepted among county taxpayers in Montana would provide revenues quite in excess of those required for necessary, though prudent, expenditures, even considering a substantial increment in the

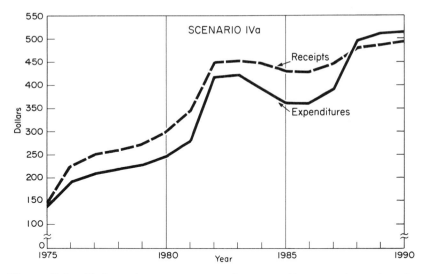

Figure 8-5. Estimated county per capita expenditures and receipts for Rosebud County, 1975–89: Scenario IVa

debt service charges to meet the amortization and interest component of current expenditures.

From all of the above, one thing stands out. It may be true that substantial advances of capital will be required to provide some facilities at the county level, but these should not be made as grants or gifts to the county government. These should be made in the form of loans, where required, or as guarantees for loans, with amortization and interest charges being financed out of the ample receipts from reasonable tax rates (as judged by Montana county standards) applied to very large tax bases. Funds available to the Coal Board out of the proceeds of the severance tax for local impact mitigation might much better be reserved for local towns which will bear the brunt of public service costs without the large mine and industrial tax base available to the counties.

CITY AND TOWN FINANCING OF COMMUNITY SERVICES

When we reviewed the financing of county government services, we found a situation in which, because of the taxable value of the mine and industrial properties contemplated, the tax base could be expected to provide adequately for standard quality community services, even at minimum tax rates. Whether this can be expected in the case of urban residents in these counties is an open question. In the case of Rosebud County, perhaps both Colstrip and Forsyth would accommodate the bulk of the migrants brought in by coal development, and while Colstrip is an unincorporated town at the time of this writing, were it to be incorporated in the future, it doubtless would have within its taxing jurisdiction the mines and related facilities that gave it its name. Forsyth, on the other hand, is some distance removed from the coal workings but would accommodate perhaps half of the population growth without having access to any of the large mine or industrial tax base that is available to Colstrip, and the county. This condition similarly characterizes Hardin, in Big Horn County, and, should Hysham in Treasure County get any of the spillover from development in the coal fields, it also would face developmental demands without having mine or industrial properties as part of its tax base.

There are three possible sources for financing essential community services for Forsyth, Hardin, and possibly Hysham. On capital account there are revenue bonds secured by the projected receipts of self-liquidating utilities, such as water and sewer systems; and also general obligation bonds secured by the "full faith and credit" of the town, namely, levies against its tax base.[10] Finally, there is the local impact account derived from the severance tax, which is available at the discretion of the Coal Board established to oversee the disposition of the severance tax receipts. Current account outlays in the first case would be financed by the charges or fees billed the customers for the utility services, while other public services such as fire and police protection, etc., would have to be financed out of general or special fund levies. Capital outlays financed out of severance taxes could be in the nature of grants or guarantees or perhaps even loans for self-liquidating investment.

In the case of self-liquidating facilities providing vendible services, the problem may not at first appear to be a significant one provided the logistics of getting facilities "on line" in time to avoid inconvenience, if not hardship, can be solved. Since these are self-liquidating facilities, providing services for which charges are made, they constitute a problem

---

[10] Legal limitations are placed on the size of the bonded indebtedness; for nonliquidating public facilities for which user charges are not practical, the total debt is limited to 5 percent of the total assessed value of the taxable property of the city or town in question. For self-liquidating facilities providing vendible services, an additional 10 percent of the assessed value of the taxable property is permitted.

not different in kind from that regularly encountered by privately managed, investor-owned public utilities,[11] or perhaps more comparably, mutual companies. These observations should not be interpreted as suggesting that the logistics of the matter are trivial—that is, getting facilities on line in time to meet demand—in circumstances where expansion, if it is to occur, will be rapid and sizable in relation to the existing base. Indeed, the terms on which financing can be arranged may be anything but favorable.[12] Uncertainties surrounding the realization of anticipated demand for such services create investor's risks that may not be overcome without some form of guarantee by higher units of government, whether the state or, if it is judged in the national interest, the federal government. It is without doubt a problem to which special attention needs to be given in the cases involving large and rapid expansion (e.g., such as that which would attend scenarios III and IV).

Facilities such as city streets, firehouses, police stations, libraries, parks, and the like also require capital outlays along with operating funds, and since these provide predominantly nonvendible public services, the facilities must be financed by general obligation bonds secured by the "full faith and credit," namely, the tax base—of the community in question. In a gradually growing community, discrete provision of these facilities can be made under conditions affording opportunity for an unhurried study of priorities. Such a gradual, well-ordered expansion would not be the case for eastern Montana localities in the event of the more rapid, intensive development of the coal fields and conversion facilities. It becomes a question then of meeting potentially explosive, but uncertain, demands on public facilities by small communities such as Forsyth, Hardin, and perhaps Hysham, with a tax base that may not justify confidence.[13] We need to review these circumstances in a quantitative manner to help gain perspective.

While ultimately not sufficient for our purposes, we begin again by reviewing the expenditure experience of communities as reflected in the

---

[11] It cannot be denied that rapid expansion of services in a relatively small town representing a small fraction of its total capacity makes it a lesser problem by an order of magnitude for an investor-owned public utility than for a small town providing its own utility services.

[12] Legal limitation on the interest allowable for payment on city and town bonds, however, is set at 9 percent, perhaps in recognition of the difficulties.

[13] As previously noted, cities and towns are limited to indebtedness for such purposes that is not to exceed 5 percent of the total assessed value of all taxable property in the city or town, *existing at the time of the last assessment*. This restriction is a constraint on the rate of expansion of capital facilities to provide community services at any given time.

most recent Census of Governments.[14] We rely on data obtained on fiscal year 1972, but as there are only eight cities in Montana possessing the minimum size to be reported in the census publication (10,000 population), we obtained comparable data from the files of the U.S. Bureau of the Census on 27 cities and towns in Montana exceeding 2,000. We sought to obtain the appropriate relation governing per capita expenditures for cities and towns by regressing observed expenditures on estimated per capita income, population size, and the per capita taxable value of property. Since we had no way of determining the size of the areas occupied by these cities and towns, we used population as an index of scale rather than population density, which in any event may be more appropriate to this case than in the case of counties. Property values were obtained from the Montana Taxpayers Association publication and were related to fiscal year 1973, that is, last half calendar year 1972 and first half 1973, which represent a data set conforming as closely as possible with the 1972 census data. The regression equation thus obtained is as follows:

$$X_1 = 36.514 + 0.00525 \ X_2 + 0.001158 \ X_3 + 0.02161 \ X_4$$
$$(0.56002) \qquad (3.6809) \qquad (0.89613)$$

$$R^2 = 0.371769 \qquad \text{S.E.} = 26.0198$$

where: X_1 = per capita town and city expenditures on noneducational tax-financed public services in dollars
X_2 = income per capita in dollars
X_3 = population
X_4 = per capita taxable value of property in dollars

The Student t statistics for the null hypothesis of no association between individual independent variables and the dependent variable are presented in the parentheses below each estimated coefficient.

Although only population size seems to be statistically significant at the 95 percent level, per capita income and value of taxable property were significant variables in some of the particular community services on which separate analyses were conducted, and thus were retained in order to maximize the explained variance. The equation suggests that about $5.25 in expenditures per capita is associated with every $1,000 rise in per capita income, $1.16 with every increase of 1,000 in population, and $21.61 per thousand dollars of per capita taxable value of

---

[14] U.S. Bureau of the Census, *1972 Census of Governments*, vol. 4, no. 3, *Finances of Municipalities and Township Governments.*

Table 8-3a. Estimated Per Capita Expenditures, Hardin, 1975–90 (dollars)

Year	Scenario I	Scenario II	Scenario III	Scenario IV	Scenario IVa
1975	106.44	106.44	106.44	106.44	106.44
1976	107.21	107.68	107.75	107.75	107.75
1977	107.35	108.29	108.84	108.85	108.85
1978	107.74	109.36	110.52	110.52	110.52
1979	108.11	109.36	110.96	110.97	110.97
1980	108.32	110.08	111.80	111.80	111.80
1981	108.67	110.23	112.47	113.31	113.31
1982	108.80	110.35	112.54	113.94	113.94
1983	109.16	110.73	113.12	117.93	117.93
1984	109.27	110.85	113.24	116.29	116.30
1985	109.64	111.23	113.42	117.31	117.31
1986	109.72	111.34	113.76	117.51	117.51
1987	109.89	111.69	113.91	117.82	117.81
1988	110.22	111.80	114.27	117.96	117.96
1989	110.32	112.16	114.41	118.30	118.30
1990	110.67	112.27	114.78	118.45	118.45

Table 8-3b. Estimated Per Capita Expenditures, Forsyth, 1975–90 (dollars)

Year	Scenario I	Scenario II	Scenario III	Scenario IV	Scenario IVa
1975	113.84	113.84	113.83	113.83	113.83
1976	114.17	114.92	115.10	115.33	115.33
1977	113.90	115.17	116.09	117.96	117.96
1978	114.49	116.13	118.10	122.16	122.16
1979	115.18	117.39	119.64	125.71	125.55
1980	115.60	118.03	121.19	128.31	126.89
1981	116.18	118.58	122.06	128.78	125.15
1982	116.50	118.84	122.74	130.45	125.11
1983	117.17	119.48	123.07	130.49	126.73
1984	117.52	119.84	123.87	128.55	129.72
1985	118.22	120.58	124.27	128.30	132.26
1986	118.57	120.95	124.89	128.44	132.41
1987	119.24	121.62	125.30	128.99	130.37
1988	119.64	122.01	126.00	129.42	129.82
1989	120.35	122.70	126.45	130.04	130.31
1990	120.76	123.12	127.17	130.45	131.92

property. Using this estimating equation, and assuming that half of the projected increase in population associated with employment increments attributable to each scenario is domiciled in Hardin and Forsyth, we obtain the per capita expenditures for the two respectively in tables 8-3a and 8-3b, on the assumption that the medium estimated per capita taxable property values ($2,000) obtain.

On the revenue side, city and town councils may levy an annual tax for general municipal and administrative purposes, and may distribute the tax receipts among such service categories as are prescribed by the ordinances of the municipalities in question. The general fund levy is limited to 24 mills, and all taxes for police, fire protection, and other general purposes for which there are no specific authorizations must come within the general fund maximum, except that an additional 5 mills is permitted if authorized by majority vote in an election in which the proposition is submitted. City and town councils have an alternative option to levy an "all purpose" rate not to exceed 65 mills excluding levies that are necessary to service bonded indebtedness or special district revolving funds. Indeed, of the 126 municipalities in Montana, 80, including a majority of the largest in the state, elect to levy an all-purpose tax. Since the practice of the single all-purpose levy is so widespread, there are insufficient observations among the special purpose functional categories even to attempt an analysis to determine the relation between their size and the size of the levy. We proceed directly to an analysis of the *total levy* as a function of a variety of variables as in the case of county government budgets.

On the assumption that the rates may vary as a function of community size, and furthermore, that communities under 2,000 population would not be directly relevant to the conditions that are postulated for the communities in eastern Montana coal counties, we selected as the sample the 32 communities with populations exceeding 2,000. A regression equation was fitted that specified the total levy to be a function of community size (population), per capita value of the tax base, and the size of the debt service levy. The resulting equation explained 25 percent of the variance but while the sign appeared correct, the per capita value of the tax base was not statistically significant. Dropping the taxable value variable gave the following equation with only marginally lesser explanatory power.

$$X_1 = 58.540 + 0.000197 \ X_2 + 1.01477 \ X_3$$
$$(1.8923) \qquad (2.5046)$$

$$R^2 = 0.231 \qquad \text{S.E.} = 8.555$$

where: X_1 = total levy in mills
X_2 = population
X_3 = special levy for debt service in mills

The values in the parentheses below the estimated coefficient are

Student t statistics for the null hypothesis of no association between each independent variable and the dependent variable.

Abstracting from the levy associated with borrowing for capital facilities, the constant along with the relatively modest influence of population variation within the relevant range resulted in an estimated total levy for all scenarios over all years for both towns, Forsyth and Hardin, of 60 mills plus or minus 1 percent.[15] Again abstracting from the millage levy that would be anticipated to service indebtedness, the levy, while not at the maximum legally permissible under the all-purpose option, would nonetheless be above the average (53 mills) for all cities and towns that opt for it.

What would such a levy applied to the estimated tax bases yield per capita? An answer to this question must be more speculative than other projections we have made. The problem, of course, is the uncertain nature of our estimates of the value of the migrants' assets that will appear as taxable real and personal property along with the nonindustrial, commercial facilities that will develop in response to the demand from the larger population. Recall that in chapter 6 we estimated crudely that the increase in real and personal property of the sort we are considering, induced by the expansion in population and commercial service facilities, would be conservatively on the order of $2,000 per capita, plus or minus 20 percent.

Given the assumption of a constant taxable value per capita, irrespective of the extent of development and the limited influence of population size, or scale effects, there would be no appreciable variation (within $\pm$ 1 percent) in the tax rate either over time, or across scenarios (except as may be required for debt service charges). In short, applying the 60 mills to $1,600 per capita taxable property as a low estimate, $2,000 as the medium, and $2,400 as the high, will give us within plus or minus 1 percent of our expected revenues. It will thus be $96, $120, and $144 for either Hardin or Forsyth, depending on which estimate of the per capita tax base one elects to use. How does this, then, compare with the tax-financed budgeted amounts per capita of other cities of Montana? The five cities with the highest per capita budget are Billings and Missoula in the over 20,000 population range with $80.27 and $75.74 per capita respectively, Kalispell in the 10,000–20,000 population range with $72.96

---

[15] This is precisely Hardin's levy exclusive of the debt service, while within the standard error of estimate for Forsyth (53.5 mills exclusive of debt service levy).

per capita, and Conrad and Glasgow in the 2,000–5,000 range with $75.05 and $71.66 per capita respectively. Moreover, these cities and towns have total levies, net of debt service charges, of 66.50, 65.11, 61.00, 60.00, and 65.11 mills respectively.

Before we take comfort from these results, however, it may be desirable to consider an additional factor, the implicit taxable value of property, per capita, in the five cities in question. If we divide the per capita tax revenue by the tax rate (including debt service), we obtain the per capita taxable value for the city or town in question. These come to $1,155, $1,058, $1,196, $1,250, and $1,048, respectively, for the cities being used for comparative purposes.[16] It is noteworthy that while these communities rank among the top ten in per capita taxable property, the value of taxable property figures to little more than half that estimated as the medium value for new Forsyth and Hardin residents and related commercial properties, and only two-thirds of the lower bound estimates for these communities. Several reasons may explain the apparent discrepancy. Either the crude methods used for estimating the per capita real and personal property described in chapter 6, although considered almost excessively conservative, in reality will be revealed to produce overly high estimates of per capita taxable value, or perhaps the property values in Montana cities are urgently in need of reassessment. The taxable value of property, primarily class V (except for automobiles), is established at 30 percent of its "true and full" value. At per capita taxable values in the $1,000 range, the value of property including home (or portion of commercial dwelling unit), vehicles, and the investments in all commercial facilities and business inventories required to service an urban population would need to be less than $4,000 per capita. This certainly is not a realistic sum, unless fixed plant and equipment in old residential and commercial establishments are computed at historic costs, and this may indeed be a factor in explaining the discrepancy. Moreover, irrespective of replacement costs, the market value of residential and commercial buildings (but not inventories) in stable, nongrowing, communities may not in fact exceed historic costs of acquisition. If this is a significant factor, the discrepancy between established values and those to be established may be more apparent than real.

There may well be a number of noneconomic factors involved that influence the assessment. Even in the absence of stable, if not declining, market value of old buildings and appurtenances in a nongrowth setting,

---

[16] *Montana Property Taxation,* 1975, p. 13 ff.

there is difficulty in carrying out piecemeal reassessment under the doctrine of horizontal equity—and it appears particularly difficult to undertake a statewide reassessment in any given year. As a result, in periods of rising costs and prices, property assessments soon become out of date and reevaluation necessary. And there may be another factor to contend with, related also to the nonexpanding Montana economy over the past couple of decades. We have in mind the work of the county assessor, which is an occult art quite apart from provisions of the law requiring "true and full" valuation. There is almost universally a community folklore or ethic regarding justice or equity in property assessment, no doubt encouraged in part by the institution of popular election of county assessors. Whatever the reasons, there doubtless exists in a period of rapid growth, or even of rapid rises in prices and costs without any real growth in demand, a discrepancy between the true and full value of the real and personal property base on which the taxable value is established and the result of assessments that have been made at some point in the not very recent past.

But there is another and potentially serious institutional complication that may affect the ability of urban communities to provide public service facilities even if the true and full value of the induced commercial and residential property were established. This is the legal limitation on the debt that a community may incur in attempting to meet capital requirements for community facilities. The amount of debt which may be incurred for facilities to provide tax-financed public services is 5 percent of the assessed value at the time of last assessment, and an additional 10 percent for self-liquidating utility (water and sewer) facilities. Typically, facilities must be expanded to meet the prospective demand of an anticipated population increment, but the assessed value base to which the amount of debt is related corresponds to the preceding year's population excluding the increment related to the assets of the current year's migrants. This matter would not be serious if the investment in facilities were modest in relation to the limitation. But this is not the case. If we take the estimated investment per capita required for providing water and sewerage facilities ($1,503 per person 1973 costs;[17] $1,650 in 1974 dollars), borrowing limited to 10 percent of the value of property as last assessed would be insufficient to meet the needed investment.

---

[17] *Anticipated Effects of Major Coal Development on Public Services, Costs and Revenues of Six Selected Counties* (Bozeman, Montana State University, 1975) p. 105, footnote 91.

This would not be serious for scenario I and II conditions, once the problem created by the steam plant construction at Colstrip in connection with units 1 and 2 (peaking before the time frame of our analysis) has been addressed, i.e., provided that past borrowing has not already utilized significantly its capacity to borrow against its tax base. A grant of roughly a half million dollars from the Coal Board, for example, would be adequate to make up the difference between the town's borrowing capacity and the required investment. The problem is virtually nonexistent for Hardin for scenarios I and II, again provided that its borrowing capacity has not been committed during past borrowing.[18] When we get into the more intensive development scenarios, particularly those that would involve conversion facilities, however, the problem would become acute in the absence of some transfer mechanism that would permit either transfers from the county (relying on its gross proceeds and industrial property tax revenues) or grants from the Coal Board to make up the deficiency between the required cumulative investment in self-liquidating utility facilities and the borrowing capacity as limited by law. The question is, would there be sufficient funds for this purpose?

There is no doubt that were the counties involved to retain a tax rate within an acceptable 45-mill bound nearer the upper level, adequate funds could be generated from the valuable new property on the tax rolls. But there is no current mechanism that permits the consolidation of fiscal resources, i.e., of the taxing jurisdictions experiencing the large new properties on their tax rolls and the urban places that are feeling the impact of social service demands occasioned by the activity on the new properties. Whether or not that can be worked out—or will gain the acquiescence of county residents who would not thereby achieve potential reductions in their tax rates, is an open question—or perhaps, on the basis of general experience, would need to be answered in the negative.

An alternative would be to use the severance tax revenues earmarked for the local impact account. Would these be sufficient to (a) meet the deficiency in borrowing capacity for the cumulative investment in utility facilities; as well as (b) care for the funding of capital outlays for school services discussed in chapter 7 in these towns? In figures 8-6 to 8-9, we display the time profile of the estimated cumulative investment, each corresponding year's borrowing capacity, and the cumulative deficit, as-

---

[18] Even were this to have been done in the past, and given the relatively modest pre-1975 borrowing capacity, a grant of a relatively modest amount in relation to the receipts of the local impact account would be a vehicle by which to deal with the problem.

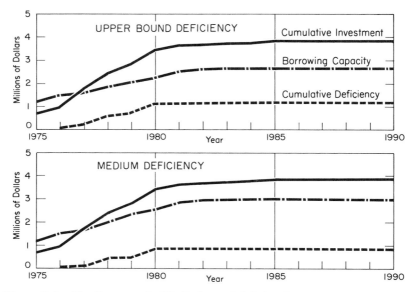

Figure 8-6. Hardin scenario III. Estimated deficiencies

suming both real and personal property taxable value per capita of $2,000 and $1,600.

Considering the problem in the case where our estimate of per capita real and personal property taxable value is $2,000, and observing the cumulative deficit between borrowing capacity and cumulative investment in utility plant and equipment for scenario III for both Forsyth and Hardin, we note that the cumulative deficit for scenario III would approximate $2 million by 1980, $2.4 million by 1985, and $3 million by 1990. When we consider that cumulative receipts by the Coal Board for local impact purposes would run in the neighborhood of $50 million, $150 million, and $255 million respectively for corresponding dates, we can rest assured that the monies would be available to meet any deficiency between cumulative investment borrowing needs and borrowing capacity for the two towns for the 100 million ton per year coal for export scenario (S-III).

What about the scenarios for more intensive coal extraction with conversion at site? While the cumulative deficiency that would need to be covered would be greater by a substantial factor ($5.0 million by 1980, and $10 million by 1985 and thereafter), we see that here as well as in the case of investment in school facilities, the receipts from the severance tax would be only modestly dented. Indeed, if we assume alternatively

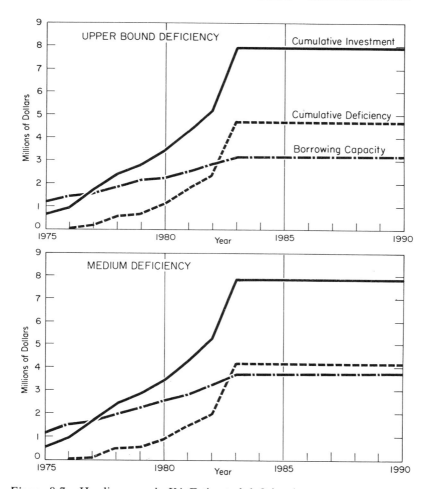

Figure 8-7. Hardin scenario IV. Estimated deficiencies

the worst case, i.e., a per capita real and personal property tax base of $1,600, and severance tax yields based on a 25 percent reduction in the contract sales price (low yield estimate of table 5-2), the peak accumulated deficit between required investment and borrowing capacity ($11.4 million) will be reached for scenario IV in 1983. The local impact account receipts, assuming the low prices for our sensitivity test, will have reached a cumulative total of $97.9 million in the corresponding year, or about eight times as much as the cumulative deficiency.

Recall, however, that we have also suggested resort to severance tax yields to meet capital outlays for educational facilities. If we look to these

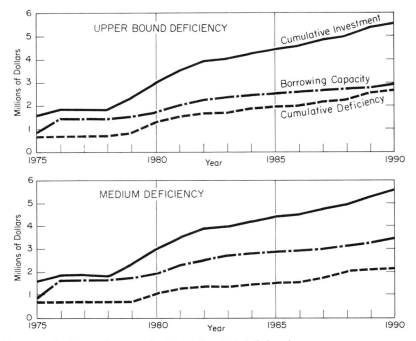

Figure 8-8. Forsyth scenario III. Estimated deficiencies

local impact fund accruals to meet the borrowing capacity deficiencies as well, we need to consider whether the combined cumulative outlays will be equal to the combined cumulative severance tax accruals for local impact mitigation. Again, if we assume that the severance taxes collected on coal taken from Big Horn and Rosebud counties are available for use by these counties,[19] and that the maximum percentage rates accrue to the local impact mitigation account, the results can be displayed most systematically as in table 8-4. Here we have the combined cumulative investment in educational facilities for the two counties, along with the combined cumulative borrowing capacity deficiency for Forsyth and Hardin, expressed as a percent of the cumulative severance tax accruals earmarked for local impact mitigation.

Taking the best estimates (medium yields and medium drafts against cumulative accruals), we find that basically after the first year (1975) accruals are equal to or exceed the capital drafts against the local impact

---

[19] This decision is essentially within the discretion of the Coal Board, thus such funds are not necessarily destined for local impact mitigation in the counties of origin.

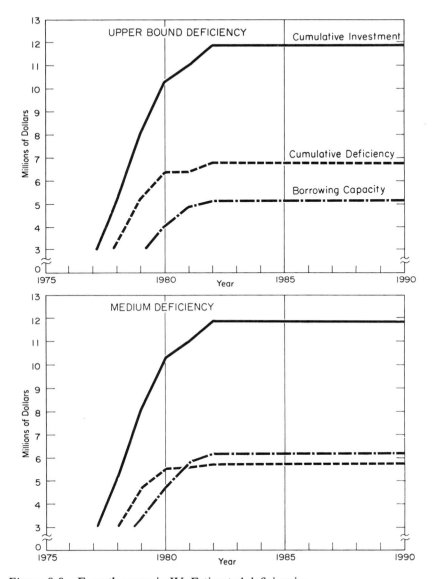

Figure 8-9. Forsyth scenario IV. Estimated deficiencies

mitigation account. Indeed, after the second or third year, it is no longer necessary to assume that all of the severance tax accruals need to be returned to the two counties of origin in order to meet the capital advances for public service facilities. This is true even when we assume lower

Table 8-4. *Cumulative Investment in School Facilities plus Municipal Borrowing Capacity Deficiencies as a Percentage of Cumulative Severance Tax Yields*

	Scenario II			Scenario III			Scenario IV			Scenario IVa		
	Lower bound	Medium bound	Upper bound	Lower bound	Medium bound	Upper bound	Lower bound	Medium bound	Upper bound	Lower bound	Medium bound	Upper bound
Low yield estimates												
1975				1.80	2.21	2.62	1.80	2.21	2.62	1.80	2.21	2.62
1976				0.82	1.00	1.15	0.84	1.03	1.22	0.84	1.03	1.22
1977				0.52	0.65	0.75	0.57	0.71	0.86	0.57	0.71	0.86
1978				0.37	0.46	0.54	0.58	0.71	0.84	0.58	0.71	0.84
1979				0.29	0.36	0.43	0.54	0.66	0.79	0.53	0.65	0.78
1980				0.25	0.32	0.39	0.48	0.60	0.71	0.45	0.55	0.66
1985				0.11	0.14	0.17	0.26	0.31	0.37	0.24	0.31	0.37
1990				0.08	0.10	0.12	0.15	0.19	0.23	0.15	0.19	0.22
Medium yield estimates												
1975	1.69	2.09	2.49	1.80	2.21	2.62	1.80	2.21	2.62	1.80	2.21	2.62
1976	0.73	0.90	1.08	0.82	1.00	1.15	0.84	1.03	1.22	0.84	1.03	1.22

	(1)	(2)	(3)	(4)	(5)	(6)	(7)	(8)	(9)	(10)	(11)	(12)
1977	0.46	0.56	0.67	0.52	0.64	0.74	0.57	0.71	0.85	0.57	0.71	0.85
1978	0.33	0.41	0.49	0.36	0.45	0.52	0.57	0.70	0.83	0.57	0.70	0.83
1979	0.25	0.31	0.38	0.27	0.34	0.41	0.51	0.63	0.75	0.51	0.62	0.74
1980	0.21	0.26	0.31	0.24	0.30	0.36	0.45	0.56	0.67	0.42	0.52	0.62
1985	0.11	0.13	0.16	0.10	0.13	0.15	0.22	0.28	0.33	0.21	0.27	0.32
1990	0.08	0.10	0.12	0.07	0.08	0.10	0.13	0.16	0.19	0.13	0.16	0.19

High yield estimates

	(1)	(2)	(3)	(4)	(5)	(6)	(7)	(8)	(9)
1975	1.80	2.21	2.62	1.80	2.21	2.62	1.80	2.21	2.62
1976	0.82	1.00	1.14	0.84	1.03	1.22	0.84	1.03	1.22
1977	0.52	0.64	0.74	0.56	0.90	0.85	0.56	0.70	0.85
1978	0.36	0.44	0.51	0.55	0.68	0.81	0.55	0.68	0.81
1979	0.26	0.33	0.40	0.49	0.61	0.72	0.48	0.60	0.71
1980	0.22	0.28	0.34	0.43	0.53	0.63	0.39	0.48	0.58
1985	0.09	0.11	0.14	0.20	0.25	0.29	0.19	0.24	0.29
1990	0.06	0.07	0.09	0.12	0.14	0.17	0.11	0.14	0.17

Note: The *severance tax* was calculated cumulatively from 1975 to 1990 assuming 17.5 percent of the total county severance tax yield (1975–79) for local impact mitigation and 15 percent thereafter until 1990. The *low yield* estimates and *high yield* estimates give a minus and plus 25 percent variation for the best medium estimate in the table. The *upper bound* combined investment and borrowing capacity deficiency assumes 300 students per 1,000 population and a low $1,600 per capita in taxable real and personal property, whereas the *lower bound* assumes 200 students per 1,000 population and $2,400 of taxable property per capita. The *medium* estimates assume 250 students per 1,000 population and $2,000 of taxable property per capita.

revenue yield and upper bound capital deficiency conditions. For example, deficiencies which would need to be met from severance tax accruals by 1980 would be only 39 percent of accumulated local impact fund accruals for scenario III, and 71 and 66 percent, respectively, for scenarios IV and IVa. By 1985, the corresponding figures for the worst cases would be 17, 37, and 37 percent, respectively, and growing progressively less with each passing year. Under our best estimate case, however, the local impact fund accruals will range from four to seven times as large as the capital required to be advanced, and an even larger multiple as time passes.

Given the ability of the severance tax receipts to remove obstacles to the necessary capital outlays for school and public utility services in the town taxing jurisdictions, it would appear likely that in every other respect the fiscal resources of the combined state and local jurisdictions, relying on the combination of the new tax base and the state severance tax, would be equal to the demands that would be made on them—given the current structure of taxes and the built-in transfer mechanisms.

CONCLUSIONS

Irrespective of the discrepancy we noted between our estimates of future assessed valuations for nonindustrial real property and those we find in established communities at the present time, the estimates of *industrial property* assessed and taxable value are likely to be close. The responsibility for assessment of such large properties is being taken over by the state of Montana, with professional auditing expertise assigned to the task. Moreover, the principal part of all coal-related taxes come from *ad valorem* taxes on the coal itself, which avoids the problem of estimating "true and full" real property value. Accordingly, whether or not our estimates of per capita real and personal property will need to yield to community custom and folkways on the residential and service-oriented commercial property, the estimates of yields from the gross proceeds and industrial property taxes are likely to remain intact. But our estimates of tax yields for city government purposes in Hardin and Forsyth will remain suspect if only because custom and practice decree otherwise. Moreover, the legal limitation on indebtedness will affect the amount of capital that can be raised for self-liquidating, revenue bond-financed facilities which are otherwise independent of the local taxing process. Depending on the claims on the local impact funds, however, this earmarked account of the severance tax could provide amply for the necessary community facilities. And it would be surprising if the local communities suffering effects from coal development in their areas would not have a superior claim to the funds in this account, which are generated by that same coal development.

In conclusion, it appears that the sources of revenues from the complex of taxes provided, and the available transfer mechanisms, whether school equalization at the county as well as the state level, or the special earmarked accounts into which revenues from the severance tax on coal are placed, could provide the funds to supply the public services demanded by the expanded population in the eastern Montana coal fields. This does not imply that the logistics of proper planning and scheduling of facilities are necessarily trivial, nor that there may not be very large risks associated with anticipatory investment to meet prospective, but not entirely certain, demands for public services. This element of risk, and the associated terms on which funds can be raised for local government services, may be the most critical problem in providing adequate public services in the coal development areas of eastern Montana—and indeed, throughout the Northern Great Plains, independent of the adequacy of the tax yields. It is clear that if coal extraction is required to meet a national energy objective, the substantial risk to the local communities' fiscal stability represents a burden that should be more equitably distributed. The question is whether the state should be responsible for assuming this risk or whether this is more appropriately a responsibility of the national government. It is clear that it is neither appropriate nor equitable to impose the risk on the local community alone.

APPENDIX C

ESTIMATION OF COUNTY INVESTMENT AND BORROWING CAPACITY

Richard E. Rice and V. Kerry Smith

The quantities plotted in figures 8–1a through 8-2b (i.e., cumulative investment, borrowing capacity, and cumulative deficiency) were determined as follows:

(1) Cumulative Investment (CI)

$$CI_j = \sum_{i=1}^{j} I_i = \sum_{i=1}^{j} K\Delta TPop_i \qquad\qquad j = 1, 15$$

where: K = estimated cost per capita for water and sewerage facilities, i.e., $1,652.80

$\Delta TPop_j$ = change in town population in year j (we assume one half the change in county population will reside in urban places)

j = 1, 15(1975–90)

(2) Borrowing Capacity (BC)

$$BC = BC_{j-1} + \alpha n \Delta TPop_{j-1} \Delta \qquad j = 2, 15$$

$$BC_1 = \alpha AV_0$$

$$\Delta = \tfrac{1}{2}[1 + \text{sign}(\Delta TPop_{j-1})]$$

and

$$\text{sign}(\Delta TPop_{j-1}) = -1 \qquad \text{if} \qquad \Delta TPop_{j-1} \leq 0$$

$$+1 \qquad \text{if} \qquad \Delta TPop_{j-1} > 0$$

where: BC_{j-1} = borrowing capacity in the previous year

α = 10 percent, i.e., the legal limit to municipal bonded indebtedness

n = estimated assessed value of real and personal property/ capita ($8,000 in the medium deficiency case and $6,400 in the upper bound deficiency case)

Δj = 1 if the change in population in the previous year is positive

= 0 if the change is negative (i.e., if the population decreases in previous year, the borrowing capacity for the current year is equal to BC_{j-1})

BC_1 = borrowing capacity in 1975

AV_0 = assessed value of property in 1974

= $8,560,000 in Rosebud County

= $11,840,000 in Big Horn County

(3) Residual Borrowing Capacity (RES)

This quantity was not plotted but is seen as the difference between cumulative investment and borrowing capacity. It is usually a negative quantity.

$$RES_j = BC_j - CI_j$$

$$= BC_{j-1} + \alpha n \Delta TPop_{j-1} - \sum_{i=1}^{j} K \Delta TPop_i$$

(4) Effective Deficiency (λ)

$$\Psi_j = \overline{RES}_j - \max(\overline{RES}_{j-1}, \ldots, \overline{RES}_1)$$

where:

$$\overline{RES}_j = abs(\overline{RES}_j)$$

if:

$$\Psi_j > 0, \qquad \lambda_j = \Psi_j$$

$$\Psi_j \leq 0, \qquad \lambda_j = 0$$

λ_j = the quantity of funds needed to meet an effective deficiency in a town's ability to finance investments in year j. That is, the magnitude of the effective deficiency

where the effective deficiency is determined by Ψ_j.

$\max(\overline{RES}_{j-1}, \ldots, \overline{RES}_1)$ = the maximum absolute magnitude of all the previous periods' residuals

The *cumulative deficiency* then is equal to the sum of the λ's through year j, or what works out to be the same thing, the maximum residual to year j.

9 Summary, Conclusions, and Policy Implications

In this study we began by reviewing briefly the energy outlook for the near-term future. If it were a matter of simply finding sufficient fixed carbon for conversion into more useful forms of energy for light, heat, and motive power, there would be little, if any, immediate concern arising out of impending scarcities, for the United States has ample stocks of carbon.[1] Fossil fuels in the form of coal are abundant, and oil shale represents an enormous potential source of hydrocarbons that can be substituted for petroleum products now obtained from more conventional sources of oil. If these potential reserves are coupled with advances in technology which open up nonconventional sources of energy through nuclear fission and fusion, the prospects for running out of energy sources appear remote indeed.[2]

But the availability of energy materials, or technologies for converting these materials into usable energy, is only one of several important considerations. Among the latter are the potential social costs of extraction, conversion, and consumption that do not appear in the conventional prices and costs registered in market transactions for energy commodities.

[1] See, for example, Sam H. Schurr and Bruce C. Netschert, *Energy in the American Economy, 1950–1975, Its History and Prospects.*

[2] Harold J. Barnett and Chandler Morse, *Scarcity and Growth, the Economics of Natural Resource Availability.*

170

These costs may take the form of diminished amenities, somatic or genetic health effects, or physical or institutional modifications of the environment that may have serious, and long-lived consequences we would want to avoid. These all represent, to members of both this and future generations, real costs which are not reflected in the current behavior of firms and individuals making market decisions concerning energy use. In the past, only part of these costs have been borne by users of energy, but if energy were priced at its full social cost, it is likely that it would be used more sparingly. Full social cost pricing would thus motivate behavior toward energy conservation, which would be an alternative, at least in part, to increasing energy supplies.

We have reviewed this possibility in summary fashion and indicated that even were this to be an effectively pursued national policy, which up to the time of this writing it has not been, the built-in inflexibility represented by sunk costs in facilities would prevent widespread introduction of energy conservation measures until these relics of an energy-affluent past could be replaced by more energy-efficient plant, equipment, and locations. In the case of structures—commercial establishments and residential buildings—and their pattern of location, the effects of past decisions may linger as much as a half century or longer before all adjustments reflecting the changes in the real relative price of energy commodities are completed. For transportation plant, or only rolling stock, for example, automobiles, it will take a minimum of a decade before the current stock is replaced completely with more efficient models and systems. Thus, even if a national policy of energy conservation were vigorously pursued, for instance, one which cut current rates of growth in half or less, there would still be a need for additional supplies of energy over the next decade or two.

Accordingly, we reviewed the sources of potential new energy supplies, assuming that there is a commitment both to full employment and to developing a measure of independence from insecure sources of oil. Looking at the available options, the ease or difficulty with which they might be brought "on line" within a decade or so, and the extramarket elements of their costs, we have concluded that increased mining of coal to substitute to a large extent for petroleum in the production of electrical energy would contribute most significantly toward a short-run solution to the energy problem. The enormous capital and environmental costs of obtaining shale oil under present technology suggest that it is not presently, nor will it in the near term, be an alternative to the more traditional source of hydrocarbons required for transportation purposes. Substitution of coal for oil in production of electrical energy is likely to release additional transportation fuels in an economically more efficient manner than would the conversion of oil shale to petroleum products. An alternative, nuclear steam electric plants, while a technical possibility for displacing fuel oil in the production of electricity, appears to offer problems in timing, if not those of another nature, while widespread application of solar or geothermal sources of energy is likely to be delayed beyond the time frame of our analysis.

Given this conclusion, we look to the Northern Great Plains coal province, namely the Fort Union Formation, as one area which has received a great deal of attention as a potential source of a large block of additional coal production. The Powder River Basin of northeastern Wyoming and southeastern Montana has been brought into production largely in response to the demands for low sulfur coal to meet air quality control standards. Potential additional demand might occur as a result of some effort toward a larger measure of energy independence and the intentions of the Canadian government to phase out natural gas and petroleum deliveries to the United States, including 50 billion cubic feet of gas per year to Montana. Moreover, if greater independence of Middle East oil is achieved by substituting coal for petroleum, and the problems of bringing in nuclear plants impede production of new electrical generating capacity in the Northwest Power Pool, Fort Union Formation coal could be looked upon as a source of energy to meet the growth in demand following the absorption by the power pool of the hydroelectric surplus that resulted from the Canadian–United States Columbia River joint development agreement. In short, there appears to be a substantial potential market for Fort Union Formation coal.

While there is a likely market for a large block of additional coal from this region, local residents are not all equally eager to see the Northern Great Plains coal beds opened up to large-scale strip mining. This appears to be particularly significant in Montana. Perhaps the environmental degradation associated with strip mining (which would not be as severe in this area as in Appalachia because of the thickness of the coal beds and moderate gradients of the land forms) may be of lesser moment in the minds of the residents of the region than the effect rapid coal development will have on the community structure, life, and style of the region. The area is characterized by a low-density, cattle-ranching economy, with few towns supporting the kind of community facilities that would be required for a population influx resulting from a large and rapid expansion of coal production, and perhaps energy conversion, in this region. It is, moreover, characterized by a stockman's subculture and life style quite different from what would doubtless replace it as a predominant way of life in the region. But these are the "currencies of change" that have been affecting rural America during the whole of the present century, whether the change occurs in an evolutionary manner that allows for gradual accommodation, or precipitously and disruptively. What has been of great concern to residents of the region has been the social disruption, if not disintegration, and malaise that has characterized the boomtown development in the Powder River Basin, especially at Gillette, Wyoming, giving rise to the term "Gillette syndrome." One important element in this phenomenon has been the inadequate provision of community facilities and services, particularly the public services that are required to maintain standards of well-being to which members of any community legitimately aspire.

The anatomy of social cohesion and well-being is perhaps the most important aspect of the "boomtown syndrome" about which we know little, and which lies within the purview of the social psychologist or perhaps cultural anthropologist who represent disciplines and competencies which have not been acquired by us. What we have sought to do, then, was to select that part of the problem to which we felt we could contribute a solution: namely, evaluating the magnitude of the fiscal issues that will be encountered if the facilities and services required to mitigate the social costs of opening up a new source of energy materials are to be provided.

A great deal of the alarm on the part of some persons and complacency on the part of others may have been fostered by the absence of useful quantitative information that would bracket the range of the impacts, and which, if known, would modify their concerns or complacencies, or at a minimum raise the informational content of the debate.

To obtain this information, we needed a method of calculating the effects of interrelated influences stemming from changes in the rate and level of coal extraction and different mixes of strategies or complexes of activities for the area that we chose for analysis—Big Horn and Rosebud counties, and the towns and school districts contained in them. We selected a recursive econometric model to simulate the pattern of adjustments to the postulated changes in regional activity associated with various levels of development. We recognize the great difficulty of using a model, which is a considerable if not gross simplification of reality (but for that reason a usable tool), to simulate accurately changes in employment, income, population, demographic characteristics, and similar attributes that provide the data for evaluating the fiscal implications of such changes in a community's structure. Under the best of circumstances, it is difficult to design a model that accurately represents reality, and when such a model is used to estimate projected changes for increasingly smaller areas, its ability to mirror reality becomes increasingly blurred. One problem here is that economic models are representations of processes played out through economic institutions—largely in more or less definable market areas—whereas the fiscal impacts in which we are interested involve uniquely political (taxing and spending) jurisdictions. Where the relevant market areas and political jurisdictions coincide exactly, the outcomes are likely to be much more accurate than where they do not, as is the case when the area of observation is progressively reduced —that is, the county, school district, and town level. However, when we know that the results are likely to be inaccurate, and we have a good idea of the direction in which our estimates depart from the real circumstances, we can use the results for judicious testing of upper and lower bound conditions to determine the sensitivity of our derived conclusions to the ranges of error that can be anticipated for different applications in different parochial circumstances. This is what we have done; namely, if a situation turns out to be not unfavorable under the worst of circumstances, it can only remain favorable under better circumstances; alternatively, if a situation is unfavorable under the best of circumstances, it can only be more unfavorable under worse circumstances. Where such outcomes will tell us all we need to know in a given case, it is foolish to reject such information in reaching judgments for which it is relevant simply because we know that our upper (lower) bound estimates are doubtless higher (lower) than the actual magnitudes.

With this in mind we used a regional econometric forecasting model developed by Curtis Harris to simulate the effects of the developments and facilities we wished to examine largely because it mimics a relatively realistic adjustment process. Because of this, it is able to track the time profile of employment and population effects resulting from the surges during construction of facilities and the declines as adjustments are made to the more permanent levels associated with the operating phases of energy production and conversion. For each of our study counties and each of five different scenarios we obtained associated employment, population, and income estimates by year, which in turn formed the basis for our estimates of the corresponding demands for public service. The model itself and the estimated economic and demographic implications of different levels and mixes of development were detailed in chapters 2 and 3.

SUMMARY OF FINDINGS

Briefly, we first took a level of 20 million tons of coal extraction for the base year 1975. Next we postulated four additional scenarios intended to reflect a representative mix of strategies and activities, varying from a modest increase in coal extraction to 42 million tons (scenario II), to a large increase consistent with a "Project Independence" level of coal extraction (scenario III). All of these scenarios, while accommodating the newly built steam electric generating plants at Colstrip, Montana, represent basically the "coal for export only" policy. That is, the policy option that would permit meeting Project Independence energy extraction goals, but would require that the coal be shipped out of the state rather than converted to another energy form within Montana. A third variant (scenario IV) assumed the same level of coal extraction (100 million tons) but made provision for the conversion of 30 million tons of coal in Montana by means of 5,200 MW of electric power capacity in Rosebud County, and a 250-million cubic feet per day coal gasification plant in Big Horn County, to represent an alternative, combined export and domestic conversion policy. This was done to permit investigating the differences in local impacts. Finally, a variant on scenario IV was proposed (S-IVa) which allowed for a more gradual time-phasing of the power plant construction to moderate somewhat the size of the employment and population surge during the construction of the power plants in Rosebud County.

Briefly, our analysis suggests that under the most extreme conditions (scenario IV) employment would about double in Big Horn County and increase by about two and a half times over the 1975 level in Rosebud County. In the latter case, however, we must appreciate that employment in Rosebud was high by historic standards in 1975 because of the power plant construction at Colstrip. Similarly, population would increase by about two-thirds in Big Horn County and by two and a quarter times in Rosebud County. The local impact differences between the policies of at-site conversion and coal for export only are quite significant. Population increases in Big Horn under the coal for export only policy would be only about a third over the base case rather than two-thirds as in the conversion scenario, while the relative reduction in population growth for Rosebud

County would be even greater, from two and a quarter times to roughly only a onefold increase.

Next, we converted the language of relevant tax legislation into appropriate computational models (chapter 4) to provide estimates of the tax yields associated with each scenario. Estimates of the yields available for state purposes were developed for each tax instrument for each scenario in chapter 5, along with the estimate of the share of royalties from coal mined on the federal lands under terms of the Mineral Leasing Act of 1920 as amended in 1976. A similar exercise was performed in chapter 6 for the local units of government.

The legislation enacted during the 44th Legislature represents a significant departure in coal tax legislation. In the first place, it moved from a tax of a given amount based on Btu content of coal alone, to an *ad valorem tax*—that is, a tax on the value of the coal itself. This was, not unintentionally, a hedge against having the real value of tax receipts eroded by inflation, and an attempt to participate in the gains from favorable changes in relative prices of energy commodities. Second, it is marked by the size of the *ad valorem* tax, that is, 30 percent of the contract sales price, for coal of the Btu content under discussion in this study. Another feature of the legislation affected the tax receipts of local governments; it changed the method of fixing the taxable value of the coal workings from a net proceeds to a gross proceeds basis, thereby greatly facilitating the administration of the tax and, incidentally, increasing its yield.

A second recent event affecting the revenues accruing to the state from coal mining on federal lands is the increase in the amount of the royalty (from 5, to not less than 12.5 percent) on coal mined from federal holdings, coupled with an increase in the state's share of the receipts from 37.5 to 50 percent. It is difficult to predict what proportion of the total projected extraction for each scenario would come from federal lands, but under some plausible working hypotheses, the change in the terms of the 1920 mineral leasing act could increase the yield by a factor of 4 to 5. Even so, the yield to the state and local governments that this would represent would still be substantially less than the yields from the severance tax (about 15 percent of severance tax yields) and combined local gross proceeds, industrial property, and real and personal property tax yields (from 55 to 95 percent of the latter, depending on the circumstances of the case).

It is clear that the overhaul of the coal tax legislation by the 44th Legislature made provision for very large increases in the tax receipts associated with coal extraction. Under plausible assumptions, the severance tax alone will yield very sizable receipts, with only that portion earmarked for local impact (15–17.5 percent) being equal to the combined receipts of all local property tax sources, including among these the proceeds determined by the gross proceeds method of valuation. This observation, of course, is subject to what conditions dominate the setting of tax rates at the local level, but as a rough order of magnitude, the severance tax yield earmarked for support of local government facilities and services will likely be equal to or exceed yields from local property taxation, given certain provisions regarding the distribution of revenues from the mandated minimum levy for the foundation educational program. These are discussed later in this chapter.

Ranked after the severance tax, mine, industrial, and real and personal property taxes, and royalties, a not insubstantial yield could be expected from the electrical producers' license tax under the more intensive (energy conversion) development scenarios. This could amount to as much as 35 percent of the yield from the state's share of royalties and somewhat more than the yield from the state tax on increased personal income associated with the increment in coal extraction and conversion, and its related employment.

Overall, the structure of Montana taxes related either directly or indirectly to coal development will yield substantial revenues for state purposes. Indeed, at the higher levels of development postulated, the coal extraction and conversion activities, under the present tax structure, would yield revenues that would represent a significant fraction of the *total revenue of the state of Montana*. It is reasonable to conclude that the fiscal impact on state government would be overwhelmingly favorable.

The question of whether the revenue–expenditure issue will be favorable or unfavorable on balance for counties, towns, and school districts will depend in part on which unit at which level of local government is under consideration. There is no question from our analysis that the yields to the counties based on the mine and industrial properties, and the methods established by law for their taxation, even at very modest rates as judged by Montana standards, will be ample to meet county government purposes. Moreover, given a countywide school equalization mechanism involving the distribution of the school district tax yields at a mandated 40-mill minimum levy, the value of mine and industrial properties would produce revenues at the 40-mill levy sufficient for meeting all of the educational needs of the school districts in the county, whether or not the properties are located within a given district's jurisdiction. But while the receipts from the properties on the tax rolls that we have postulated would be more than enough to finance the educational services of the county system, the proceeds of the mandated 40-mill minimum levy are restricted to providing only the foundation-level educational program, with all proceeds surplus to this expenditure level being destined for the educational equalization program at the state level (see chapter 7). Ironically, there appears to be no net advantage to the county school districts (except for the one with the new large taxable property on its rolls, i.e., a smaller additional rate will suffice to support the permissive and voted enrichment programs), since the state equalization mechanism would guarantee the foundation-level program in the absence of the new developments. This observation is not made to raise any latent issue of equity in the school equalization program, which we believe to serve worthy social objectives. It is made only to suggest that the affected school districts—principally the ones in the towns that are called upon to service families of workmen at mine and industrial facilities located outside their immediate jurisdictions—may have a superior claim to the earmarked local impact funds generated by the severance tax from coal extraction in these counties. This is a significant point since the earmarked local impact funds are not automatically destined for grants to the counties of origin.

If we review the demand for educational services as implied by the population and demographic characteristics associated with each scenario (chapter 7), and the investment in plant and facilities to accommodate these demands, the cumulative investment at any time after the initial year or two will represent a decreasing fraction of the cumulative severance tax proceeds earmarked for mitigating local impacts. Depending on other competing claims of equal merit, the severance tax yields earmarked for such local impact purposes appear to be comfortably ample, irrespective of scenario, county, or other related factors, such as the sensitivity of results to the particular working assumption adopted where judgment was required in the absence of very accurate data (see tables 7-3a and 7-3b). In drawing this conclusion, it is worth emphasizing that costs of facilities were based, not on historic, but on replacement costs of standard educational facilities, all computed in terms of 1974 dollars. That there may be an inflation in costs beyond 1974 is not doubted, but we have expressed contract sales prices and other elements relevant to assessed valuation in constant dollars as well so that changes in the general price level as distinct from changes in relative prices are of no moment in evaluating our conclusions.

The bulk of the tax-financed costs would be occasioned by the need to provide educational services, and thus our analysis here was based on incremental costs using architectural data. With only about 23 percent of the remaining local government tax-financed costs representing expenditures for county purposes, we did not feel it necessary to carry out an equally detailed study of each of the several public functions performed by county governments. But, since Montana counties range in size from several hundred persons up to nearly a hundred thousand, a cross-sectional analysis of tax-financed expenditures was run using data from the 1972 Census of Governments to determine the relationship between per capita expenditures, and income, community size, value of taxable property, and area of county. Similarly, since expenditures are not basically independent of receipts in budgeting for public expenditures, we also attempted to estimate the feasible expenditure budget indirectly by regressing the tax rate on the same set of variables, and applied these rates to the projected value of taxable property to obtain indirectly an equivalent measure of the revenue–expenditure profile for the two eastern Montana counties, as the postulated scenarios were played out. As we suspected, the values of the two sets of estimates (expenditures and receipts, per capita) were very close for corresponding scenarios, counties, and years. Indeed, since the budget for tax-financed local public services is prepared against the background of anticipated tax receipts, in reality they need to be close whether or not our estimating equations would have produced such results. In any event, since the two methods provide somewhat different means of estimating the feasible tax-financed expenditures for various circumstances characterized by different populations, income, value of taxable property, and so on, corresponding to very different levels and character of developments over the years, we have at least a way to approach the problem of evaluating the fiscal impacts of different events for the area under review.

A comparison of the per capita expenditure–revenue estimates generated by our equations with the current expenditure–revenue experience within the state may be instructive. At rather modest tax rates, estimated as a decreasing function of the value of the tax base in part, the revenues produced for county government use would place these counties initially within the top quintile of all Montana counties and eventually, for the more intensive development scenarios, into the 90th percentile. Indeed, for Rosebud County, in the more intensive development scenarios we could visualize feasible per capita expenditure budgets six to ten times as large as counties within the same projected size range, with Big Horn three to five times the average.[3] All of this suggests that because of the relation of yields, even accounting for declining tax rates related to increasing taxable property values that would accompany the large-scale expansion of coal extraction and conversion, the county fiscal situation would appear to be very favorable.

We conclude this in spite of the knowledge that the cross-sectional analysis is based on a single year's data—and therefore to a large extent, past investment will represent a part of the costs (debt service) and receipts (value of taxable property in sluggish real property market areas). Thus these estimates will reflect the historic costs—for the bulk of the plant and equipment represented by our data. While this is true, it is nonetheless the case that our equations contain the relevant variables the values of which are expected to change dramatically in a manner reflected in our projection model, and while the coefficients may not remain stable over time, we feel our results warrant confidence by virtue of our cross referencing with the results of the analysis in chapter 7. That analysis addressed incremental new investments in standard educational facilities valued at 1974 construction costs. The results there indicated that even at the mandated 40-mill minimum levy, the yields would be sufficient to cover all costs, debt service figured conservatively, as well as operating costs. There is no reason to feel that current costs of the other public service facilities would differ in this respect. Thus, given the available flexibility in the tax rate while still remaining within the acceptable range as judged by Montana local government experience, there is no reason to feel tax yield in the coal counties would fall short of prudent public service expenditures.

Perhaps this is a point that should be emphasized for its implications regarding policy with respect to the financing of county government services. The tax base and hence yields for any reasonable tax rate will be so large in relation to prudent public service expenditures that there will be no need for grants or transfers from other state funds, for example, the severance tax local impact account, *for county government purposes.* The relationship between tax base and public services for the counties is very much more favorable than for the towns in the county that will have to meet the increased demands for public services without having access to the tax yields from the large mine and industrial properties that give

---

[3] Extreme as this may appear, the estimates are consistent with experience in Powder River County, which has similarly high extractive resource values.

rise to these demands. It is conceivable, but not likely, that for some purposes capital advances might be needed, but these should be treated in the manner of loans for repayment rather than grants. There appears to be no reason why grants or transfers of funds should be made from other jurisdictions, given the affluent fiscal circumstances that would develop for the counties in the coal area.

However, the circumstances affecting the towns, as well as those affecting the school districts in the towns where the public services will be demanded, will suffer from the dissociation of the taxing jurisdictions in which the taxable mine and industrial properties occur from those on which the demand for services will in large part fall. As a consequence, it follows that the towns, as well as the urban school districts, will be in much poorer circumstances than the counties for supplying tax-financed services. But here is where our analysis is the weakest. Our estimates of the incremental value of taxable real and personal property associated with the urban population increment corresponding to the various scenarios rests on very shaky evidence. On the one hand, our estimates are quite conservative compared with those of knowledgeable persons "on the ground" so to speak. On the other hand, the estimates of $2,000 of taxable real and personal property per capita (plus or minus 25 percent) appear much higher than the official assessed valuation of real and personal property in existing towns and cities in Montana. Several factors could account for this apparent discrepancy. Again, it may be simply the difference between the bulk of the taxable properties in other Montana cities and towns being acquired in the past and valued largely at historic rather than current costs and values, whereas the incremental value of real and personal property associated with the population increment used in our study would be valued at current costs and market values. Part of the problem may be the need for statewide reassessment of property because there has been no updating of assessed values in the recent past. If these were the reasons, then we would not need to be concerned about the apparent discrepancy. On the other hand, the problem could be that the county assessors have an unwritten "code of equity" that operates in the background of their assessment processes quite independently of the legal injunction to assess "the true and full" value. It is a truism that whenever assessed and market values of real estate are compared, there is universally a marked discrepancy. Accordingly, we are not certain whether our estimate of the taxable real and personal property of new migrants, in any event done in rather cursory fashion, is a good estimate of the true and full value and will be assessed in like manner, or whether the community folkways governing the local assessment process will produce different results which would vitiate our analysis of the fiscal resources available to the towns in relation to their need to provide public services.

If our estimates are correct and the discrepancy is due to the difference between the market value of real estate in stable or declining communities where the market is inactive (consistent with Montana experience generally speaking over the past quarter century), and market values at times exceeding current replacement costs where the market locally is hyperactive, tax yields might suffice to meet public services, although we are skeptical even about this. If the estimates of taxable property have not reflected "on the ground" assessment procedures, and we have greatly overestimated the likely assessed values, then it is quite clear that the financing of local services would run into problems. Again, referring to our analysis of the revenues for supporting local educational expenditures in chapter 7, we observed that when we took only the real and personal property tax base (at our estimated valuation), and even at a 120-mill tax rate, the receipts were insufficient to cover expenditures. If other public services at incremental current costs were roughly consistent in this respect with the current costs of providing educational services, we would not expect the towns and cities to fare well in the fiscal context.

Second, even when we address the question of self-liquidating, vendible utility services that would not depend on taxes but on revenue bonds for their financing, we encounter a problem related to the legal limitation on borrowing expressed as a percentage of the existing property valuation at last assessment. Investments in public utility facilities are required in advance of the demand for the services, whereas the base which limits the borrowing represents a past assessment. This may be no problem for stable communities looking toward replacement of facilities, but it is a problem of great potential severity in rapidly expanding communities such as those which have the more intensive development scenarios in our analysis. However, if we analyze the problem in terms of the estimated capital investment, cumulative over the years, with the corresponding community borrowing authority, we can estimate the deficiency that would need to be provided. Doing this, we learn that the cumulative grants that could be made by the Coal Board to cover the deficiency represent a relatively modest claim against the cumulative local impact funds. Indeed, if we assume that the capital outlays *for affected school districts* and *the deficiency between investment and borrowing capacity for public utility facilities* were both provided for by the Coal Board from the local impact fund, the combined cumulative total would represent after the initial year or two a decreasing fraction of the local impact account accruals (see table 8-4 for results and sensitivity tests).[4] Accordingly, while we are somewhat skeptical regarding the ability of the towns and urban communities in the coal area that do not have valuable mine and industrial properties to finance the resulting demand for public services imposed on them, the receipts available to the Coal Board for mitigating local effects of coal development and related activities appear quite ample for meeting possible deficiencies.

POLICY IMPLICATIONS

Applicability of Study Methods and Results

Now, what can be said about the broader policy implications that arise out of our analysis? A background observation is in order before we address the national policy implications as well as those related more narrowly to state and local governments. It should be obvious that while this study has been devoted to the intricacies of state and local fiscal issues for the state of Montana and two of its eastern coal counties, its implications are of much wider significance. Among other things, we are speaking of a level of coal expansion which in the limit represents a quarter of the total increment in production for the nation proposed by the recent (spring 1977) presidential energy program. But aside from this, we are dealing with an area of state, local, and by implication federal, fiscal relations that has been surfacing in much the same form, not only in other western coal-producing states, but also in coastal states facing a rapid buildup of infrastructure to exploit offshore oil occurrences. The opposition of these states, and the potentially affected communities, to oil refineries on the eastern seaboard, and the political and legal action taken by the states in connection with the federal government's Outer Continental Shelf leasing program attests to the widespread concern about "boomtown" development. Unfortunately, the web of laws and administrative rulings—the institutional milieu—will vary from one state to the next, since it is the product of independent legislatures, administrative boards, and policies which evolve within each state in response to perhaps unique political circumstances. This study of the problem of fiscal impacts within the institutional setting of Montana law and administrative policy is then

---

[4] This is based on accruals to the local impact account at the maximum stipulated 17.5 percent of severance tax yields 1975–79, and 15 percent thereafter.

in a sense unique to Montana, but it addresses a problem which is surfacing in much the same form, in different settings, in connection with other large-scale energy and mineral developments. Both methods and results should therefore be of wider interest and, it is hoped, of direct assistance to Montana policy makers.

Severance Taxes and Conservation

Having reassured ourselves that the study is not merely parochial, let us move immediately to an implication that is not jurisdiction-specific. We shall return ultimately to issues specifically related to state and local fiscal structures. In chapter 1 we concluded that the growing dependence on Middle East oil, which ironically has increased dramatically since the embargo of 1973, suggests the importance of conservation as one element of a national energy policy that would provide a larger measure of independence from a predominant and uncertain area of supply. We also concluded that subsidies for energy production and use, including a failure to charge for the environmental side effects, have led to a less than optimal amount of conservation in the U.S. economy. A clear implication, at the federal level certainly, and perhaps also at lower levels, is that the costly and often hidden subsidies ought to be reduced or eliminated, and the market failures corrected.

In principle, the way to promote efficiency in energy use—and energy conservation—is to ensure that the price to ultimate users reflects its true social cost. As a general observation, subject to more discriminating review below, instruments of taxation are useful means by which to make the price reflect the external costs not included in conventional market prices. The use of the severance tax as a device to raise revenue from the extraction of coal in order to defray the infrastructure—and, perhaps also the environmental—costs of its development represents a better method of mitigating local impacts, from the point of view of national economic efficiency, than a grant-in-aid from the federal government or any other source of general taxation. And as we also suggest later on, it has some equity considerations in its favor as well. In this way, the social costs of providing the associated services occasioned by coal extraction are included in the price of the energy commodity to guide market behavior of energy users.[5] We have seen that, given the possibility of a rapid and ex-

---

[5] This requires a relatively inelastic demand for the commodity in question, a condition that holds in this case. [See Robert B. Shelton and William E. Morgan, "Resource Taxation, Tax Exportation, and Regional Energy Policies," *Natural Resources Journal* vol. 17, no. 2 (April 1977).]

tensive expansion of mining activities and their associated infrastructure costs, sizable advances of capital might be needed by towns for facilities to provide necessary public services. These advances may be in addition to the funds available to local jurisdictions, whether through taxation of local property or borrowing secured by the prospective revenues from sale of vendible services. Such grants, however, if made from the proceeds of the severance taxes on the coal itself, would be reflected in the market prices of the commodities, and would provide an appropriate incentive to more efficient energy utilization. This serves economic efficiency, and perhaps equity, goals as well as the national interest in energy conservation. The alternative of grants from general tax revenues, whether from the federal government or elsewhere, will lead simply to a perpetuation of the national policy of subsidizing consumption, which is responsible to a significant extent for the present energy problem.

Of course, just as it is important to ensure that prices of energy commodities reflect the full social costs so that energy will not be used for purposes in which the benefits as measured by the price at the margin are less than the costs, so also is it important that the prices of energy commodities do not exceed the true social costs. There have been some criticisms of the size of the Montana and North Dakota severance taxes, and our own results suggest that a relatively small fraction of the severance tax receipts would meet the fiscal (as distinct from psychic) burden of local impact mitigation. The matter has even been presented as a potential constitutional issue on the grounds that the size of the tax, allegedly in excess of the need to cover costs, represents a monopoly power which, unless the demand for Northern Great Plains coal is not perfectly price inelastic, would represent a state-imposed restraint on trade.[6] That is, the pattern of energy commodity flows in interstate commerce would be altered, and from an economic point of view, distorted from the efficient spatial location of extraction, transportation, and conversion. We note this here in passing, and will return to it later on in the discussion of state and local policy implications.

A justification for the size of the severance tax sometimes advanced by those representing the local or regional interest relates to the fact that in all likelihood the use of fossil fuels represents only a transitory phase while advanced, unconventional sources and technologies are developed. Accordingly, a rapid buildup of extractive and conversion activities in

---

[6] Richard Nehring and Benjamin Zycher with Joseph Wharton, *Coal Development and Government Regulation in the Northern Great Plains: A Preliminary Report.*

the region can be expected over a relatively short time span, when viewed from a larger societal perspective, followed by an abrupt decline, leaving in its wake depleted coal deposits and a potential twentieth-century "ghost region," or in any event a seriously depressed economy with attendant social problems. To compensate for this, it is argued that it is desirable for the state that will inherit these problems to prepare to cope with them by establishing a trust fund out of the proceeds of the sale of the coal from which the rest of the nation benefits, and which exhausts the natural assets of the state.[7] Opponents of this position argue that this represents an assertion of property rights to resources which are in part on lands owned by the federal government.

Whether or not the "taking" issue can be advanced successfully by opponents of the alleged confiscatory severance tax rate, the assertion of property rights raises a related issue. There are various environmental amenities that attract, or retain, some elements of the population, even when other social, cultural, and economic amenities are absent, and as a result give evidence of their value to residents with these preferences. These amenities, which are basically common property resources, are destined to be adversely affected by any large-scale coal extraction, transportation, and/or conversion activities. The degradation of the environment will represent real income or welfare losses to residents, as well as a diminution of option value for perhaps prospective future residents of the region. These are real costs that will not appear in any market-determined price and must be reflected by addition to the price, which the severance tax accomplishes, whether or not a mechanism exists for compensating the losers out of such tax revenues.

Severance Taxes and Economic Efficiency

The positions that have been advanced in the controversy over the intergovernmental fiscal relationships which we have attempted to summarize above may now be addressed more analytically.[8] As in most discussions

---

[7] Interestingly, the state during a referendum in the autumn of 1976 voted to allocate half of the proceeds of the severance tax for this purpose.

[8] All of the subsequent analysis is based on the presumption that because of the volume of coal reserves in this region, a permanent solution to the energy and related issues will have been developed through some "backstop technology" that will displace coal as a source of energy substantially before rising prices induced by impending depletion become relevant for our purpose here. In this event, we can abstract from the dynamic considerations introduced by Hotelling's concept of "user costs." Now, to the extent that rising prices might be induced for coal due to the progressive depletion of alternative energy sources over the period of our analysis, we consider these likely to be of the second order of smalls. Accordingly,

of economic policy, there are two basic issues: efficiency and equity. The former includes the conservation of energy. The latter involves both members of the communities affected by development and the consumers of energy commodities in the wider market area, as well as other certain intergenerational effects. A third issue of some political, if not economic, concern, has to do with what political jurisdiction is the appropriate one for dealing with these problems.

In considering the efficiency question, we need first to observe that the size of the severance tax is critical. It is known to be larger than the rate necessary to cover the infrastructure costs, but other elements of social costs have not been specifically addressed, either in this study or by others. These include the health and aesthetic effects of air and water pollution, the psychic costs of transition for residents of the region, or the displacement of these residents and the loss of future options.

We do not know of course whether the difference between the severance tax yield and the cost of providing the interdependent investment in public service facilities (a difference of some 80 to 85 percent) is just equal to the omitted elements of social cost, but a little tax incidence analysis can reveal the efficiency implications of this possibility and the alternative—that the tax exceeds the omitted elements.

Case 1. We assume, first, that the tax just equals relevant external costs. Further, we assume that the supply of coal up to the levels covered by our analyses can be provided at constant real unit extraction costs, which is plausible because of the enormous reserves. Finally, we assume that the demand for eastern Montana coal is relatively inelastic because of the unavailability of adequate substitutes. In other words, even a substantial increase in price will not lead (over the period of our study) to much of a decrease in quantity demanded. This is also a plausible assumption for at least three reasons: (a) much of the current coal mined and contracted for is being used to displace higher sulfur coals used outside the region to meet national air quality standards; (b) there is a need to find replacement energy for the 50 billion cubic feet of natural gas per year imported into Montana from Canada, and which is now being phased out by the Canadian government; and (c) there is an impending deficit between estimated load growth in the Pacific Northwest and timely al-

---

we consider it unnecessary to complicate the analysis for our purposes with these refinements. [For a review of the problem when user costs are considered to be of practical significance, see Stuart H. Burness, "On the Taxation of Nonreplenishable Natural Resources," *Journal of Environmental Economics and Management* vol. 3, no. 4 (December, 1976) pp. 289–311.]

ternative energy sources to satisfy the demand. Given these assumptions, then, the theory of tax incidence leads us to expect that the severance tax would be shifted forward to the ultimate consumers, who would equate its value in use at the margin with the full social cost. The result would be completely consistent with national economic efficiency. If we wish to assume for further extension of the argument that the demand elasticity is given by the estimates made by the Federal Energy Administration,[9] we would expect that the quantity demanded would be reduced by about 15 percent compared with that in the absence of the severance tax. Again, assuming the tax rate (30 percent) is a good first approximation of the difference between the private and social costs of energy extraction, conversion, and transportation, not only would the results be consistent with national economic efficiency, they would also be consistent with a measure of energy conservation.

Case 2. Now let us assume that the conditions previously assumed hold in every respect except that the tax rate exceeds the difference between private and social costs. What efficiency effects would we anticipate? Obviously the ultimate cost to consumers will exceed the true social cost, and the amount of energy consumed will be less than the efficient amount. Two things might be said about this. We are dealing in a "second best" world and it behooves us to consider whether the resulting inefficiency is greater or less than that associated with the most likely alternative. One such alternative would be a grant-in-aid from the federal government to underwrite local infrastructure costs, subsidizing perhaps part of the private resource cost as well as ignoring the various elements of external cost we have noted. Another alternative would be simply to adjust the severance tax rate downward to conform more closely to the external social costs omitted in private market transactions.

Accordingly, in the case in which we have inelastic demand and constant-cost supply conditions, we conclude that if the severance tax rate is correct, at least to a first approximation, the tax serves both efficiency and conservation objectives without qualification. When these supply and demand conditions exist, but the tax rate exceeds the amount of the uncompensated elements of social cost, the matter of efficiency is problematical. While the tax rate can be judged inefficient by Pareto criteria, it might also be judged with reference to the alternative that would prevail. There is not much reason to expect that the federally sponsored

[9] Federal Energy Administration, *National Energy Outlook,* appendix C (Washington, GPO, 1965).

alternative, the subsidy, would result in a solution closer to the Pareto optimum. Indeed, since at the federal level so many issues other than efficiency appear to have intruded, one may doubt that an alternative "second best" solution would necessarily serve the interests either of efficiency or of conservation.

Case 3. Finally, let us assume that coal from this area cannot be provided at constant supply price for the quantities postulated in our more intensive extraction scenarios. What will be the likely effects of the severance tax? The results, again drawing on the theory of tax incidence, would be that only part of the tax would be shifted to the consumer, with part falling on the owners of the coal deposits. Here, if the part shifted onto the consumer were just equal to the difference between marginal private and social costs, then the Pareto criteria would be met. If not, prices to consumers would be either higher or lower than required for Pareto-efficient outcomes, but not necessarily worse than would occur from a plausible federally sponsored alternative. In one respect, however, the outcome would differ—there would be income and wealth transfers from the energy resource owners to the beneficiaries of the severance tax. But this is a question of income *distribution,* which leads us to the discussion of the second class of issues, those involving equity.

Severance Taxes and Equity

As always, it is much more difficult to say anything very conclusive about the equity aspects of a policy. Accordingly we shall need to be even more diffident here than we were in discussing the efficiency issue.

The basic issue is, who bears the burden of the social costs associated with the coal development—resource owners (whether private or common property),[10] consumers, or general taxpayers? Let us assume, for the sake of argument, that the ethically preferred outcome is the one involving the most progressive income redistribution—that is, the one in which the costs fall most heavily on those most able to bear them. Unfortunately, we don't really know which of the groups concerned—re-

---

[10] There are psychic costs that are likely to be incurred by nonenergy-resource owning residents for which there is no compensation mechanism. This is a troublesome social and ethical problem that needs to be recognized even though it cannot be dealt with satisfactorily. In short, the earnings from the tax justified by such costs are not used to compensate adversely affected third parties suffering "common property resource" losses.

For a more extended discussion of equity considerations, see Allen Vander Mullen, Jr., and Orman H. Paananen, "Select Welfare Implications of Rapid Energy-Related Development Impact," *Natural Resources Journal* vol. 17, no. 2 (April, 1977).

source owners, users, or general taxpayers—in general are best off, and which are worst off, in terms of predevelopment income. This problem is a good candidate for further policy analysis. In any event, for the moment we can trace out in a rough, qualitative way the incidence on each group corresponding to each of the cases discussed above.

In case 1, which involves a severance tax equal to external costs, and constant return to scale, the costs are borne largely by the users of the energy products.[11] No doubt some would argue that this is equitable, in a sense. Those enjoying the benefits bear the costs. But with respect to our suggested criterion that the costs fall most heavily on those most able to bear them, again we must concede that we don't know whether this means the users as a class.

Alternatively, financing the local public service costs by grants-in-aid from the federal treasury would shift the burden to the general taxpayer. But as economists are quick to point out, the subsidy of primary commodities or intermediate goods and services (here energy commodities and electricity) is a very ineffectual way of achieving income redistribution in the ethically correct direction, whatever it may be. Indeed, like the rain, the public largesse tends to fall on rich and poor alike under these circumstances, and it is not clear that the ultimate redistribution is not perverse. For example, equity owners in the resource extractive industries may gain a windfall if location characteristics that affect regional demand elasticities permit pricing energy commodities in private markets above their long-run marginal extraction and transportation costs. Although it is conceivable that monopoly pricing may be consistent with pricing to reflect full social costs, and would thus represent a move in the direction of energy efficiency, the justifications for subsidizing production on grounds of equity are specious since the redistributive effects may well be perverse.

A final point to consider here relative to third-party effects is that, to the extent the severance tax is included in the price of the factor and final consumption service, and the demand for Fort Union Formation coal is less than perfectly inelastic, it will have the effect of reducing the extent of development and thus affect a smaller number of persons than would a policy whose results were to subsidize production, with some part of the subsidy passed forward to purchasers of factor and final consumption services. If anything, the equity argument relative to third-party

---

[11] Ignoring the uncompensated third-party costs reflected in the diminution of the area's preexisting services of common property resources.

effects then is in favor of the severance tax compared with the assumed alternative.

Now let us consider case 2, in which the severance tax exceeds the difference between private and social costs of energy development. What then would be the equity implications of the severance tax? If, as indicated by the conditions assumed, the tax would be passed on to coal and other energy buyers in factor and final consumption service markets, there would be a transfer of income from consumers of the coal and converted products to the state of Montana and to the beneficiaries of the expenditures of such funds from the various earmarked accounts. It is beyond the scope of this study to trace through some representative sets of distributional effects under the variety of plausible conditions. It would be appropriate to surmise, however, that under some plausible assumptions the redistributive consequences would be inconsistent with the prevailing social ethic. On the other hand, when they are compared with the consequences of the plausible alternative in a second best world—federal subsidies to local units of government in lieu of severance taxes—it is not clear which scheme is preferable. In either case, it appears that a fourth group—residents of the coal region—would benefit at the expense of the energy users or the federal taxpayers. For a region suffering two decades of economic stagnation, during which it had fallen from the top quarter of per capita income to the lowest third, there may be something to be said for this on grounds of distributive justice. Of course, the question remains as to whether the implied redistribution ought to come from users or taxpayers. Without a great deal of further analysis of this problem, which is quite outside the scope of this study, one would be inclined to consider the jury as being out on this issue.

Finally, let us again consider case 3, in which the long-run marginal cost of coal from the Fort Union Formation is an increasing function of volume of output in the upper range postulated in our study. What then might we expect in terms of distributional consequences? With a rising long-run marginal cost of coal extraction, the incidence of the severance tax would fall on both the consumers of the coal and converted products, on the one hand, and on the owners of the resources to which the severance tax applied on the other. The degree to which the incidence would be shifted forward to the consumer, or back on the owner would be determined by the elasticities of the supply and demand relationships and the ranges over which the analysis was relevant. All that can be said here, however, is that returns (economic rents) to owners would be reduced, and where federal lands would be at issue, the receipts of the federal government in bonus bids would be affected. In short, there would be a transfer of income, or wealth, from the beneficiaries of the economic rents from federal coal deposits, whoever they may be, to the beneficiaries of the Montana state severance tax. Moreover, the less elastic the supply of Fort Union Formation coal, the greater would be the degree of transfer of economic rents from federal (and other) coal lands to the state. Further unraveling of the pattern of transfer, and ethical judgment on it, are beyond our inclination, if not competence.

The Question of Optimal Jurisdiction: Federal, State, or Local

Discussion of federal versus state expenditures and receipts does, however, bring back into focus the "third issue" noted earlier: Which unit of government, federal, state, or local, is appropriate for dealing with the first two issues of efficiency and equity? We look briefly at two aspects of the jurisdiction issue here: federal versus state administration of the severance tax, and large local unit (such as county) versus small (such as school district) for the disbursement of earmarked severance tax receipts, or collection and disbursement of local property taxes.

Taking the second first, and assuming for the time being that the constitutionality of the severance tax is not at issue, the question of the jurisdiction in which the tax should be collected and disbursed to cover infrastructure costs warrants comment. It appears that some mechanism might be considered that would set aside revenues from the tax base of a county for use of the residents of a smaller jurisdiction who suffer adverse impacts from the activities of large industrial properties on the county's tax rolls. This could avoid the problem of the county experiencing fiscal opulence with which it has difficulty coping while its urban communities, suffering the public service demands without commensurate tax revenues, are straining to meet these demands. Some mechanism which would consolidate the fiscal resources of town and county, or town and larger area into a special multicounty coal development and local impact mitigation district merits consideration.[12] It could provide a wholesome environment in which those who are the most immediately affected by the events and their consequences would have the responsibility for raising and disbursing public funds.[13] This would avoid the situation that

---

[12] Where there are interstate externalities, as between the activity in Big Horn County, Montana, and the dormitory town of Sheridan, Wyoming, the problem is politically much more difficult, if not impossible. Though the mechanism of an interstate compact is available, getting the Montana legislature to accede may be a very difficult matter indeed.

[13] See Harold M. Groves, *Financing Government.*

naturally arises with the present mechanism, the local impact account of the state severance tax. Here incentives exist for county residents to press for lower tax rates on their properties so that the large industrial properties finance county government, while urban communities are encouraged to press for excessive claims against the local impact account for perhaps imprudent investment, in the knowledge that this account is open to claims from other sections of the state that are not affected by developments in the coal region. Substituting the federal government for the state in dealing with local communities only increases the incentives for self-serving strategies.

Indeed, the county of a special district might play a more effective fiscal role, using its taxing and licensing authority to fix the costs of social overhead on the activities most directly responsible for such costs. An example is the mechanism of the county axle-weight license fee as a means of raising revenues for the maintenance of roads used by, or for, the transport of heavy equipment in the coal region. And as we have already suggested, it would be consistent with national economic efficiency and the conservation of energy to fix the charge specifically on the economic agents involved in the energy extraction–conversion process who give rise to the public costs, in lieu of transfers from one or another public revenue source that is not specifically related to the activity inflicting the costs. Moreover, this is consistent with some notions of equity.

Turning now to the other jurisdictional issue, namely, whether imposing the severance tax as an instrument of policy to regulate use and bring the price of energy commodities into equality with their full social cost is a function that should be assumed by the federal government or by a state, we can offer only a few casual observations.

The basic question is whether a state can impose taxes and regulations having energy and environmental implications in a manner consistent with the broader national viewpoint. Those who pose the question do so with the conviction that states cannot. Theoretically, national interests can be addressed adequately only by national assemblies. No state legislature, it is asserted, will be governed by the larger national interest when its constituents' welfare would otherwise be adversely affected. While this is doubtless true in large part, it is equally argued by critics of congressional behavior that state delegations are also more responsive to their own than to the general public's perception of the national interest. That there are many who feel so is manifested by the fact that in a year in which the country witnessed both its president and vice president removed from office for impeachable, not to say criminal, offenses, the public opinion polls nonetheless continued to reflect greater confidence in the White House than in Capitol Hill. Accordingly, the assertion that state initiative in these areas represents the "Balkanization" of national policy must be weighed against some well-motivated questions arising out of past congressional conduct as the appropriate guide for policy in the public interest.[14]

If we were to discover that the Montana state severance tax on coal was sufficient, to a workable approximation, to cover the difference between the costs which are mediated through market transactions, and the full social costs (or if it was found excessive, reduced to the appropriate level), we might as readily argue that the action of the 44th Montana Legislature represented an example of efficient decentralized decision-making rather than an example of Balkanization. Indeed, to give a flavor of how the matter was viewed in some quarters at Helena, the following quote from a high-level staffer early in 1975 is revealing:

> If the feds think we are going to take our marching orders from Washington when we see them milling around in confusion, they don't have the wit to recognize the national interest.

However else the action of the government of Montana may be judged, it appears that by proceeding independently it has developed a self-reliant energy policy and was prepared to perform well before its federal counterpart worked its way through what seems to many an interminable political thicket. Since there is a public benefit from making a large source of coal available in timely fashion to meet the demands of a national energy program, some credit is due Montana for moving more decisively in an area not without its own political thickets.

This is an issue of inherent complexity and because of our own lack of expertise in this area, the posing of the issue may be as much as should be required of this study. Without benefit of a careful, sophisticated analysis of these problems, the view of Montana's contribution toward resolving the near-term energy problem is likely to depend on which end of the telescope the viewer is privileged to look through. One thing is certain in this area of uncertainty, however. The value of the contribution will be viewed differently if seen from the perspective of Pennsylvania Avenue than from Last Chance Gulch.

---

[14] For a vigorous defense of the role the states can play, see, for example, Ira Sharkansky, *The Maligned States: Policy Accomplishments, Problems, and Opportunities.*

Bibliography

Almon, Clopper. *The American Economy to 1975* (New York, Harper & Row, 1966).

Barnett, Harold J., and Chandler Morse. *Scarcity and Growth, the Economics of National Resource Availability* (Baltimore, Johns Hopkins University Press for Resources for the Future, 1963).

Baumol, William J. "Macro-Economics of Unbalanced Growth: The Anatomy of Urban Crisis," *American Economic Review* vol. 57, no. 3 (June 1967) pp. 415–426.

Bender, Lloyd D., and Robert I. Coltrane, "Ancillary Employment Multipliers for the Northern Great Plains Province." A report to Economic Research Service of U.S.D.A. and presented to the Joint Meetings of the Western Agricultural Economics Research Council's Committees on National Resource Development and Community and Human Resource Development, Reno, Nevada, January 7–9, 1975.

Bjornstad, D. J. *Fiscal Impacts Associated with Power Reactor Siting: A Paired State Case Study*. A report supported by the U.S. Nuclear Regulatory Commission (Oak Ridge National Laboratory, Oak Ridge, Tennessee, September 1975).

Bureau of Reclamation and the Institute of Applied Research, Montana State University. *Anticipated Effects of Major Coal Development on Public Services, Costs and Revenues in Six Selected Counties* (Bozeman, Montana State University, 1975).

Burness, H. Stuart, "On the Taxation of Nonreplenishable Natural Resources," *Journal of Environmental Economics and Management* vol. 3, no. 4 (December 1976) pp. 289–311.

195

Carlsmith, R. S., R. L. Spore, and E. A. Nephew, "Systems Studies of Coal Progress Report—December 31, 1974." Work supported by NSF Inter-agency Agreement No. AEC 40-418-93 and NSF AG 398 (Oak Ridge National Laboratory, Oak Ridge, Tennessee, February 1975).

Council on Environmental Quality. OCS Oil and Gas—An Environmental Assessment (Washington, 1974).

Darmstadter, Joel, and Hans Landsberg. "The Economic Background of the Oil Crisis," Daedalus vol. 104, no. 4 (Fall 1975) pp. 15–37.

———, with Perry D. Teitelbaum, and Jaroslov G. Polach. Energy in the World Economy: A Statistical Review of Trends in Output, Trade, and Consumption Since 1975 (Baltimore, Johns Hopkins University Press for Resources for the Future, 1972).

Division of Research and Information System, Montana Department of In-tergovernmental Relations. "County Profiles" for Yellowstone, Treasure, Powder River, and Big Horn Counties (April 1974).

Duncan, Donald, and Vernon Swanson. "Organic-Rich Shale of the United States and World Land Areas," U.S. Geological Survey Circular 523 (Washington, GPO, 1965).

Federal Energy Administration, National Energy Outlook, Appendix C. (Washington, GPO, 1976).

———. Project Independence Report. (Washington, GPO, 1974).

Ford Foundation. A Time to Choose: America's Energy Future. A final re-port prepared by the Energy Policy Project of the Ford Foundation (Cam-bridge, Mass., Ballinger, 1974).

Groves, Harold M. Financing Government (5th ed., New York, Holt, Rine-hart and Winston 1960).

Hammond, Allen L., William D. Metz, and Thomas H. Maugh II. Energy and the Future (Washington, American Association for the Advancement of Science, 1973).

Harris, Curtis C., and Frank E. Hopkins, Locational Analysis (Lexington, Mass., Lexington Books, 1973).

———. The Urban Economies, 1985: A Multiregional, Multi-Industry Fore-casting Model (Lexington, Mass., Lexington Books, 1973).

Isard, Walter. An Introduction to Regional Science (Englewood Cliffs, N.J., Prentice-Hall, 1975).

Kohrs, Eldean V. "Social Consequences of Boom Growth in Wyoming." A paper given at the Rocky Mountain American Association of the Advance-ment of Science Meeting, April 24–26, 1974, Laramie, Wyoming.

Krutilla, John V., and Anthony C. Fisher. The Economics of Natural Envi-ronments (Baltimore, Johns Hopkins University Press for Resources for the Future, 1975).

Leholm, Arlen G., Larry F. Leistritz, and Thor A. Hertsgaard. "Fiscal Im-pact of a New Industry in a Rural Area: A Coal Gasification Plant in Western North Dakota." Paper for presentation at the Seventh Annual Meeting, Mid-Continent Section, Regional Science Association, Duluth, Minnesota, June, 1975.

———, Larry F. Leistritz, and Thor A. Hertsgaard. Local Impacts of Energy

Resource Development in the Northern Great Plains. A report prepared for NGPRP (Department of Agriculture Economics, North Dakota State University, Fargo, North Dakota, September 1974).

Montana Department of Intergovernmental Relations, Research and Information Systems Division. *The Economic Impact of Proposed Colstrip Units 3 and 4 on the Rosebud County Economy* (Helena, Montana, August 1974).

Montana Energy Advisory Panel. *Coal Development Information Packet* (State of Montana, Office of the Lieutenant Governor, December 1974).

Montana Taxpayers Association. *Montana Property Taxation*. Editions for 1973 and 1975 (Helena, Montana).

Montana Taxpayers Association. *Property Tax Budget Guide* (Helena, Montana, 1975).

Mountain West Research. *Economic Impact Assessment Summary Report*. A report prepared for Westmoreland Resources with respect to coal development options for Tract II and Tract III Crow Indian lease areas, Big Horn County, Montana (Tempe, Arizona, 1976).

Mountain West Research. *Environmental Baseline Studies for Crow Indian Coal Leases, Socio-Economic Report*, 3 parts (Tempe, Arizona, 1975).

Muller, Thomas. *Fiscal Impacts of Land Development, A Critique of Methods and Review of Issues* (Washington, The Urban Institute, URI 98000, 1975).

Nehring, Richard and Benjamin Zycher, with Joseph Whorton. *Coal Development and Government Regulation in the Northern Great Plains: A Preliminary Report*. Prepared in part under a grant from the National Science Foundation (Santa Monica, Rand, 1976).

National Academy of Sciences. *Rehabilitation Potential of Western Coal Lands*. A report to the Energy Policy Project of the Ford Foundation (Cambridge, Mass., Ballinger, 1974).

National Petroleum Council. *U.S. Energy Outlook, Nuclear Energy Availability*. A report on the Nuclear Task Group of the Other Energy Resources Subcommittee of NPC's Committee on U.S. Energy Outlook (Washington, NPC, 1973).

Nordhaus, William D. "Economic Growth and Climate: The Carbon Dioxide Problem," *The American Economic Review, Papers and Proceedings of the Eighty-Ninth Annual Meeting of the American Economic Association*, February 1977.

Northern Great Plains Resources Program. *Draft Report* (Denver, Colorado, September 1974).

Northern Great Plains Resources Program. *Effects of Coal Development in the Northern Great Plains, A Review of Major Issues and Consequences at Different Rates of Development* (Denver, Colorado, April 1975).

Northern Great Plains Resources Program. *Socio-Economic and Cultural Aspects Work Group Report* (June 1974).

Nuclear News (February 1974).

Office of the Superintendent of Public Instruction. *1975 Montana Schools Statistics* (Helena, Montana, 1975).

Polzin, Paul E. *Projections of Economic Development Associated with Coal-Related Activity in Montana.* A report prepared by the Bureau of Business and Economic Research, University of Montana (Missoula, Montana, January 1974).

Polzin, Paul E. *Water Use and Coal Development in Eastern Montana* (Montana University Joint Water Resources Research Center, Bozeman, Montana, November 1974).

Power, Thomas M., John W. Duffield, John R. McBride, Richard L. Stroup, Terry D. Wheeling, William D. Thomlinson, Walter J. Thurman, and Arnold J. Silverman. "Montana University Coal Demand Study, Final Report, Projections of Northern Great Plains Coal Mining and Energy Conversion Development 1975–2000 A.D." (Research support from NSF/RANN, November 1975.)

Reports of the State Department of Revenue (Helena, Montana, for the periods July 1, 1970 to June 30, 1972; June 1, 1972 to June 30, 1974; and July 1, 1972 to June 30, 1975).

Schurr, Sam H., and Bruce Netschert. *Energy in the American Economy, 1950–1975, Its History and Prospects* (Baltimore, Johns Hopkins University Press for Resources for the Future, 1960).

Scott, Claudia Devita. *Forecasting Local Government Spending* (Washington, The Urban Institute, 1972).

Sharkansky, Ira. *The Maligned States: Policy Accomplishments, Problems, and Opportunities* (New York, McGraw-Hill, 1972).

Shelton, Robert B., and William E. Morgan, "Resource Taxation, Tax Exportation and Regional Energy Policies," *Natural Resources Journal* vol. 17, no. 2 (April 1977).

Spore, R. L., E. A. Nephew, and W. W. Lin, "The Costs of Coal Surface Mining and Reclamation: A Process Analysis Approach." Paper presented at the Western Economic Association 50th Annual Conference, San Diego, California, June 25, 1975. Research sponsored by NSF/RANN program under Union Carbide Corporation contract with ERDA—Energy Division, Oak Ridge National Laboratory, Oak Ridge, Tennessee, 1974.

State Commission on Local Government. "Financial Statement for Cities, Towns and Counties." Fiscal Years 1972–1974. (Unpublished reports provided by Steve Turkiewicz, SCLG, Helena, Montana.)

State of Montana Department of Highways Planning and Research Bureau. *Analysis of Contracts Awarded, Calendar Year 1975* (Montana, 1975).

U.S. Atomic Energy Commission, Division of Reactor Research and Development. *Power Plant Capital Costs, Current Trends and Sensitivity to Economic Parameters* (Washington, GPO, October 1974).

U.S. Bureau of the Census. *1972 Census of Governments,* vol. 4, *Government Finances,* nos. 3 and 4 (Washington, GPO, 1974).

U.S. Congress, Senate. Committee on Interior and Insular Affairs, *Geothermal Energy in the U.S.,* Committee Print 92-31, 92 Cong. 2 sess. (Washington, GPO, May 1972).

U.S. Department of Commerce, Regional Economics Division, Bureau of

Economic Analysis. *Economic Profiles of the Northern Great Plains Region* (U.S. Department of Commerce, 1973).

U.S. Department of the Interior. *Final Environmental Statement for the Prototype Oil Shale Leasing Program,* vol. 1, *Regional Impacts of Oil Shale* (Washington, GPO, 1973).

U.S. Department of the Interior. *Prospects for Oil Shale Development: Colorado, Utah, Wyoming* (Washington, 1964).

U.S. Department of the Interior, Geological Survey. *Final Environmental Statement Proposed, Plan of Mining and Reclamation, Big Sky Mine, Peabody Coal Company, Coal Lease M-15965, Colstrip, Montana, FFS 74-12.* 2 vols. (March 1974).

Vander Mullen, Allen, Jr., and Orman H. Paananen, "Select Welfare Implications of Rapid Energy-Related Development Impact," *Natural Resources Journal* vol. 17, no. 2 (April 1977).

Weichman, Benjamin. "Energy and Environmental Impact from the Development of Oil Shale and Associated Minerals." A paper presented at the 65th Annual Meeting of the American Institute of Chemical Engineers, November 1972, revised 1974.

Willrich, Mason, and Theodore Taylor. *Nuclear Theft: Risks and Safeguards* (Cambridge, Mass., Ballinger, 1974).

Index

Library of Congress Cataloging in Publication Data
Krutilla, John V
    Economic and fiscal impacts of coal development.
    1. Coal—Montana—Case studies. 2. Power resources
—Great Plains—Case studies. 3. Energy policy—Great
Plains—Case studies. 4. Great Plains—Economic
conditions—Case studies. I. Fisher, Anthony C., joint
author. II. Rice, Richard E., joint author.
III. Resources for the Future. IV. Title.
HD9547.M6K78   333.7   77-89300
ISBN 0-8018-2054-5